1984

THE
ROBOTICS
PRIMER

Robert A. Ullrich is a professor of management and associate dean of the Owen Graduate School of Management at Vanderbilt University in Nashville, Tennessee. He serves as a member of the Education and Training Committee of Robotics International of the Society of Manufacturing Engineers, and he has written several books, including *Motivation Methods That Work* (Prentice-Hall/Spectrum Books, 1981).

Robert A. Ullrich

THE
ROBOTICS
PRIMER

The What, Why, and How
of Robots in the Workplace

A SPECTRUM BOOK

Prentice-Hall, Inc., Englewood Cliffs, New Jersey 07632

Library of Congress Cataloging in Publication Data
Ullrich, Robert A.
    The robotics primer.

"A Spectrum Book"
Bibliography: p.
Includes index.
1. Robotics. 2. Robots, Industrial. I. Title.
TJ211.U44   1983   629.8'92   83-17805
ISBN 0-13-782144-1
ISBN 0-13-782136-0 (pbk.)

This book is available at a special discount when ordered in bulk quantities.
Contact Prentice-Hall, Inc., General Publishing Division, Special Sales, Englewood
Cliffs, N.J. 07632.

A SPECTRUM BOOK

1   2   3   4   5   6   7   8   9   10

Printed in the United States of America.

ISBN 0-13-782144-1

ISBN 0-13-782136-0 {PBK.}

Prentice-Hall International, Inc. *London*
Prentice-Hall of Australia Pty. Limited, *Sydney*
Prentice-Hall of Canada Inc., *Toronto*
Prentice-Hall of India Private Limited, *New Delhi*
Prentice-Hall of Japan, Inc., *Tokyo*
Prentice-Hall of Southeast Asia Pte. Ltd., *Singapore*
Whitehall Books Limited, *Wellington, New Zealand*
Editora Prentice-Hall do Brasil Ltda., *Rio de Janeiro*

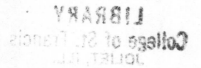

/

For Honey

CONTENTS

Preface

WHAT IS A ROBOT?

According to a definition adopted by the Robot Institute of America, a robot is a reprogrammable, multifunctional manipulator designed to move materials, parts, tools, or specialized devices through variable programmed motions in the performance of a variety of tasks.

The key words here are *multifunctional* and *reprogrammable*. Unlike conventional automation, robots are designed to perform, within design limits, an unspecified number of different tasks. Robots are equipped to perform one task or another by programs that are created by the organizations that use them.

Modern robots are controlled by microcomputers, but the programming of these computers is unconventional. For example, in the case of continuous path robots, programming takes the form of leading the robot by its hand (or end *effector*, as this part of the machine is called) through the sequence of steps that will comprise the task the robot is to perform. Thus, when programmed by a highly skilled worker, the robot will mimic the worker's skills. Robots with continuous path control allow the user to specify the path that the robot will follow in moving from one point to another. Machines with point-to-point control, alternatively, allow the user to specify only the points between which the robot will move—the path taken between these points will be determined by the robot.

Robots can be equipped to sense heat, pressure, electrical impulses and objects, and can be used in conjuction with rudimentary vision systems. Thus, they can monitor the work they perform. To put it another way, the modern robot can learn and remember tasks, respond to its work environment, operate other machines, and communicate with plant personnel when malfunctions occur. It represents a new technology—one that is predicted to reshape the way we think and work.

THE
ROBOTICS
PRIMER

1

An Introduction

THE FAYETTEVILLE ROBOTS

We are accustomed to thinking of this country as a sprawling network of cities. Yet most of our land still lies in farms, prairies, and woodlands that have not appreciably changed from one century to the next. The towns that dot these vast rural areas have the timeless quality of the land, and they are slow to shed the customs of the last century for those of the present. Such places are well off the major highways; they attract few visitors and little attention.

Fayetteville, Tennessee, is such a place. South of Nashville the ridges and deep hollows of the Tennessee Valley gradually subside into lush, wooded hills and gentle, rolling farm land. State roads meander among these fields and pastures. Where the roads come together, towns such as Fayetteville have grown.

If one took away the automobiles that line its streets, Fayetteville, Tennessee, could be used as a location for a motion picture set in the turn of the century. Retirees still gather on benches in the town square to voice their old men's thoughts. The buildings that enclose the square are of another, earlier era. People on the sidewalks are unhurried, friendly, and slow and thoughtful in their speech. A few miles outside Fayetteville, on the 231 Bypass, a small brick factory sits atop a hill. Inside this factory, robots assemble electric motors.

Actually, the factory manufactures oscillating electric fans. It is hardly an exaggeration to say that sheet metal, wire, and plastic go into one end of the plant and completed fans emerge from the other. A half dozen of the fans' components, such as screws, are manufactured by suppliers. Everything else—electric motors, fan blades, pedestals, and the like—is made in the factory from raw materials.

Tennessee Fan Company is a wholly owned subsidiary of its parent company, Matsushita Electric Ind. Co., a Japanese firm. In addition to the fact that its employees work side by side with robots, this company is unusual in other respects. The factory and its products were jointly designed in an attempt to achieve both high product quality and maximum plant productivity. Another interesting point is that the presence of laborers among the robots and other automated machines satisfies political requirements rather than the economic objectives of the firm. One of the conditions required of Tennessee Fan Company in return for permission to operate in

Tennessee stated that the firm must create employment opportunities. Today the firm employs 85 individuals, including 15 technicians and a plant manager—about half the number of workers required to produce the same volume of similar products using more conventional manufacturing methods. However, with the exception of the plant manager and the technicians, the firm's employees perform tasks that essentially are no different from those performed by the firm's robots. In fact, had the firm not been constrained in its use of robots and related devices, little human labor would be required. Using equipment available today, the company could operate with only the 15 technicians and the plant manager. In Japan, other factories function similarly now. American plants of this type are on the drawing boards.

The juxtaposition of these robots and their small-town environment is striking, but it probably would be a mistake to make too much of it. It would be unwise to think that the good people of Fayetteville are awed by the little machines that assemble motors or that their lives will be much changed by them. To find the future in a setting that typifies our rural past is interesting, but ultimately it is the future that should stimulate this interest rather than the unusual way in which we observe it.

And yet there is a lesson to be drawn from the town and its small robot population: We must gain the wisdom to know when to change—and what not to change. We have come dangerously close to losing our traditions of hard work, decency, and personal integrity during the past decades. Fortunately, they have survived more or less intact, particularly in rural America. Our small towns have retained other traditions that are worth preserving as well, but other things must change if we are to be successful in meeting the challenges of the next decade. The ways in which we work and think about working must change. The schooling that prepares young adults to lead productive lives must adapt, in part, in anticipation of fundamental changes in production methods. Many of the traditional attitudes and customs that have served us well in Fayetteville, Boston, and Los Angeles must be discarded before we can prosper again as a nation. Our attitudes toward business risk, competition (particularly from abroad), taxation, government regulation, applied research in industry, technological innovation, and myriad other factors affecting national productivity are in need of review and reassessment.

or on the surfaces of other planets. Similar work is being focused on robots that will repair satellites in orbit. Finally, according to Dr. Georg Von Tiesenhausen, of NASA's Marshall Space Flight Center, we may be only 20 years from the creation of a robot that can reproduce itself from scratch.[4]

Twenty years is not a very long time. Can we accomplish such astounding technical and scientific advances in just two decades? The old men who congregate in Fayetteville's square might tell you that we can. You see, in their own lifetimes, these men have witnessed technological advances from gas lights to space flights. They can remember when the Wright Brothers flew at Kitty Hawk. They have also seen the Mars lander robot search for life on a distant planet.

Be that as it may, we must concern ourselves with the present, for the robot revolution in industry has just begun. It is imperative that we understand and benefit from this new technology, if for no other reason than to remain competitive with other industrial nations.

This book seeks to provide a description of robot technology and its present and future uses for managers and interested lay persons. Beyond this, the book explores the managerial and social issues with which robot labor ultimately will confront us. The following chapter explains the principle of robotics in nontechnical terms and contrasts robotics with conventional automation. In Chapter 3, we take a more detailed look at the ways in which robots work. Chapter 4 examines the economics of robot labor and looks at the question of when it is to a firm's advantage to replace human workers with robots. Chapter 5 looks into the research laboratories where the capabilities of tomorrow's robots are being developed. Chapter 6 addresses the impact that tomorrow's technology is likely to have on the management and structure of organizations. Current evidence suggests that labor relations and organization structure, as well as day-to-day management practices, must adapt to the advent of this new technology. Finally, Chapter 7 looks to the future—to the social consequences of a robot economy.

To date, relatively little has been published in this new but growing field. Much of this small body of literature is technical and does not provide an adequate introduction to the subject for managers and other readers who are not technically trained. Nonetheless, several excellent books are available, and I recommend them most highly.

---

[4]Bob Dunnavant, "Self Reproduction by Robots Forecast," *The Tennessean*. Friday, March 27, 1981, p. 14.

For these reasons, I have provided an annotated bibliography of the robotics literature for those readers whose interests are stimulated and who seek information beyond the scope of this book.

As I said, the scope of this book is introductory. It is high time, then, that you were introduced to the world's population of 30,000 robots.

2

What Is a Robot?

A BREAKTHROUGH IN AUTOMATION

The company signs that have begun to appear in industrial parks and office complexes give one the feeling of having overslept about 20 years and awakened, like Rip Van Winkle, in the future—signs that say *Advanced Robotics Corporation*, *Cybotech*, *Machine Intelligence Corporation*, and *United States Robots, Inc.* Machine *what?!!*

Once inside these factory gates, though, a feeling of the present returns, for these companies produce equipment that does not appear to be markedly different from the automated machines with which we have long been familiar.

Automation is not a recent innovation. By the end of the eighteenth century, for example, the Frenchman Joseph Marie Jaquard had introduced automated looms that were controlled by punch cards similar to the familiar IBM card. Twelve years after their introduction over 11,000 such automated looms were operating in France alone. Before another 25 years had passed, Charles Babbage had constructed his Difference Engine, a forerunner of the modern computer.

By the beginning of the present century, automation had taken hold in our factories, and well before World War II automated production lines, such as those found in bottling plants, were commonplace. Today, automation is not only commonplace, it is taken for granted: the temperature in our homes, the tuning of our FM receivers, and the speed of our cars can be automatically controlled for us. We are accustomed to automated machines, and robots tend to look to us like more of the same. It would be a mistake, though, to think that they are.

Traditional automated systems are designed and constructed to do one, and only one, thing. They are efficient and effective at what they do, but they lack versatility. For example, a bottling plant cannot be used to build electric motors. Moreover, even apparently simple changes such as, in the case of the bottling plant, a change in the shape of the bottles to be filled usually require elaborate, expensive alterations of the production system. Such changes literally require that portions of the system be redesigned and rebuilt.

Robots are different. For one thing, they can be designed to perform a variety of different tasks. They are versatile machines, having a wide range of applications. For another, they can be designed to check their work, make decisions, and alter their performance accordingly.

ROBOT DEFINED

The essential differences between robots and conventional, hard automation are contained in a definition that was adopted by the Robot Institute of America. The definition states that a robot is ". . . a re-programmable, multifunctional manipulator designed to move material, parts, tools, or specialized devices through variable programmed motions for the performance of a variety of tasks." The key words here are *multifunctional* and *reprogrammable*. Unlike hard automation, which is designed to perform a single function (say, filling pop bottles), a robot is designed to perform a variety of functions that are not specified in advance by the designer. Much like a human hand, which is versatile enough to play a violin and to wield a sledgehammer, an industrial robot in theory could be designed to perform tasks as dissimilar as these are. Now, having hands does not imply having the ability to play in the string section of an orchestra or to use a 12-pound hammer. Obviously, both skills must be learned. Similar to people, robots have the ability to learn different skills. They can be programmed, in the sense that a computer is programmed, to perform specific tasks, and they can be reprogrammed to perform new tasks.

I have used the foregoing analogies between robots and people advisedly, for the term *robot* can evoke visions of humanoid machines, such as C3PO, that are far removed from the realities of industrial robotics. Today's robots neither look nor act like people. Tomorrow's probably will not either.

INTRODUCTION TO A ROBOT

There is no compelling reason to build an industrial machine that can do everything that you and I can do. Imagine, for instance, an assembly line worker who is an accomplished violinist in her spare time. It would be sheer folly to attempt to replace such a worker, or even an average worker, with machines having all the talents those workers possess, even if we could do so.

For this reason, industrial robots are designed to perform a very small number of the tasks for which we humans are suited. Thus, the analogy between the marvelously dextrous human hand and the robot hand is extremely crude, as is the analogy between human learning and robot programming.

Yet, robots are fairly talented in their own right. Take T³⁽ᴿ⁾, for example. T³ is a versatile automaton built by Cincinnati Milacron. As you can see from the accompanying line drawing of this machine, T³ looks less like C3PO than it does something one might find in a dentist's office. As I said, robots do not appear markedly different from the machines to which we have grown accustomed.

However, this one machine can perform a relatively large variety of tasks without the aid of an operator. For example, it can arc weld, load and unload machine tools, perform machine tool operations, and conduct quality control checks. How it performs these operations is a topic that is addressed in the next chapter. For the present, I would like to review the anatomy of this robot.

As illustrated in Figure 2–1, T³ is a system of three separate integrated machines: a power supply that can be likened to the human circulatory system; a microcomputer that is analogous to the brain; and the robot itself, which, if you like, comprises the robot's skeletal and muscular systems. You may have noticed that the robot's arm does not have a hand: it ends in a circular plate. The details of the hand were omitted from the drawing because robots such as T³ can be equipped with a variety of different hands, each uniquely suited for a particular kind of work.

FIGURE 2–1 T³⁽ᴿ⁾ Computer-Controlled Industrial Robot (Illustration courtesy of Cincinnati Milacron)

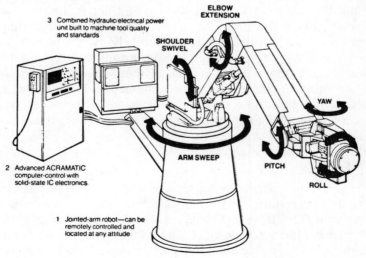

Robot Hands

The typical robot hand is designed to do relatively few things: for example, to grasp, to turn, and perhaps to sense and limit the force it exerts on what it holds. Rather than construct a hand capable of playing the violin, if such were possible, manufacturers design robot hands for relatively simple tasks because it is more economical to do this, and to replace a robot's hand when the nature of the robot's task changes, than it would be to design and build a general-purpose hand for the robot.

Robot hands come in a variety of styles, none of which remotely resembles a human hand. A robot that handles sheet metal or plate glass may have a row of automatically operated vacuum cups for a hand. If this robot is reassigned to the task of loading small parts on a pallet, it may be refitted with a two-fingered hand that looks like a pair of tongs. Alternatively, a welding torch, grinding wheel, or tool bit may be attached in place of the hand. There are as many different types of hands as there are ingenious solutions to the problem of putting robots to work.

As this and some of the discussion that follows will show, economic considerations usually limit the versatility that can be designed into industrial robots using current technology. Robots are not designed as general-purpose machines, although in theory they could be designed as such. Rather, they are built to perform restricted classes of tasks. Just as an *end effector*, as robot hands are called, that is designed to lift plate glass is not suited for manipulating small parts, a robot that can move parts weighing several hundred pounds may not be suited, by virtue of its cost and design limitations, for removing minute castings from a die. Similarly, a robot that can follow the intricate curves of an automobile body may be well suited for spray painting but may not be an appropriate machine for removing bricks from a kiln. Yet, all of these robots, by definition, must be multifunctional and reprogrammable. That is, the spray painting robot that has a spray gun for an end effector can be programmed to paint Volkswagens, Ford trucks, or unicycles.

Robot Arms

It is difficult to imagine being able to work with only one arm; it seems that everything we do requires two. Yet, if you were strong enough and skillful, you probably could do many things with one hand tied behind your back. You probably use your left hand just to guide and steady your right hand when using tools such as screwdrivers and electric drills. Robots are steadier than you and I, and more accurate when performing repetitive tasks. Thus, most of them need only a single arm.

The arm illustrated in Figure 2–1 is designed to move in a manner similar to that of the human arm. The arm sweep and shoulder swivel approximate the range of motion of the human shoulder. However, two joints are required to achieve this motion in the robot, whereas the human shoulder consists of a single ball and socket joint. The robot's elbow extension joint is comparable to the human elbow. The wrist is more complicated, and three separate joints are used in the robot to achieve a range of motion roughly comparable to that of the human wrist.

Each of the distinct types of motion that the robot arm can make is known as a *degree of freedom*. T^3 is designed to move with six degrees of freedom; each degree of freedom is illustrated by one of the arrows in Figure 2–1. Just as end effectors, or hands, are selected according to the type of task to be performed by the robot, the number of degrees of freedom designed into the robot will limit the kinds of work it can do. In general, the greater the number of degrees of freedom, the more versatile the robot. However, each degree of freedom adds to the cost of the robot. Thus, less expensive machines, having relatively few degrees of freedom, ordinarily are preferred for tasks that do not require the use of more complex machines.

The range of motion that can be achieved by T^3 is illustrated in Figure 2–2. The surface of the volume through which the arm can move describes the robot's maximum reach. The robot can reach a point on this surface from only one direction. Points that lie within the volume, though, can be reached from more than one direction, and the number of directions from which such points can be reached is a function of the number of degrees of freedom possessed by the robot. In certain tasks, the robot may be required to reach around the object on which it is working in order to work on the object's back or side. Obviously, such tasks require robots with more than a few degrees of freedom.

Robot Legs

What is the maximum range of motion of the human arm? One could say that at present this range is defined by a sphere having the earth at its center and a radius of approximately 240,000 miles—roughly the distance between the earth and moon—because human arms have done work on the moon.

The analogy to the robot is quite simple: The robot's range

FIGURE 2–2 A Robot's Range of Motion: The volume of space through which the end effector of the T³ robot can move. (Illustration courtesy of Cincinnati Milacron)

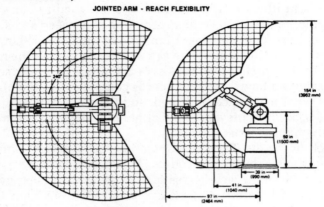

JOINTED ARM - REACH FLEXIBILITY

of motion will be expanded to the extent that we can enable the robot to move about.

Experimental machines have been built with legs. They have the virtue, at least in theory, of being able to negotiate uneven, unpredictable terrain. For the industrial robot, though, nothing as elaborate as mechanical legs is needed. In robotics, as in most engineering endeavors, the elegant solution is preferred.

An industrial robot may need the ability to move about a factory. To achieve this merely requires the addition of a set of rails on which the robot can travel. These may run along the factory floor and permit the robot to move alongside an assembly line or to travel between machine tools. One of the robot's virtues, though, is its relative insensitivity to its working conditions. Thus, it is possible to hang the robot by its heels, so to speak, from something on the order of an overhead crane. Suspended like an acrobat, the robot will be able to drop in on a number of work stations that it could not otherwise reach.

SPECIES OF ROBOTS

What has been described so far is the robot: The rest—the power supply and controller—is peripheral. Before going on to introduce these peripheral machines, I would like to touch briefly on the variety of robot types that are available, for the discussion so far merely has described the outward appearance of a single robot. The inner complexities of robots are described in advanced technical publications, some of which are listed in the bibliography. Such technical detail is beyond the scope of this book. However, a brief introduction to the subject is essential to the following discussion of peripheral equipment.

In describing the volume of space through which T^3 can move nothing has been said about how the robot actually moves from one point in the volume to another. This is an important aspect of robots. Some robots, when programmed to move from point A to point B, do not permit specification of the path that the robot's arm will travel to get from A to B. This is acceptable for the performance of some tasks but not for others, such as certain spray painting jobs, where it is important to keep the spray nozzle a certain distance away from the surface being painted. In order to spray paint automobile bodies, for example, it must be possible to control the path of the robot as it moves from headlight to taillight in order to keep the spray nozzle a fixed distance from the curving surface of the car body. T^3 has point-to-point control. The programmer merely specifies the sequence of points that the robot is to reach, and the robot moves between each pair of points in a straight line. Other machines, called *continuous path robots*, permit the programmer to specify both the points to be reached and the path to be followed between each pair of points.

Next, there is the matter of the speed with which the robot travels between points. In some cases, it is sufficient merely to have the robot move as quickly as possible between points. In others, such as in the case of a robot that is used to weld seams along compound curves, it is necessary to control precisely not only the robot's path but also its rate of travel along the path.

Precision of motion is another important attribute of robots. The degree of accuracy achieved by the robot in returning to each point in the sequence of points that comprise a particular task will be more or less important depending on the nature of the task being performed. T^3 can return time and again to a point in space with an accuracy of ± 0.05 inches, and this margin of error can be halved by adding a special option to the robot.

Precision, or repeatability, is a quality that must be paid for by the purchaser. Greater or lesser precision is available in other types of robots. The degree of precision achieved by T^3 may not be demanded of a robot that is used, for example, to unload bricks from a kiln.

Robots also vary in strength. T³, for example, can lift up to 225 pounds. At the other extreme, the little Microbot Minimover −5^{(R)}, with five degrees of freedom, can manage an eight-ounce load with a precision of ±0.013 inches. Again, the lifting capacity of a robot must be matched to the requirements of the tasks for which the robot is purchased. T³ is a general-purpose industrial machine. Minimover −5^{(R)} is designed for educational and experimental purposes and is priced to be affordable to the serious hobbyist.

ROBOT CONTROL

As you might suspect, robots also vary according to the type of control systems they employ. The simplest robots are controlled by sets of movable switches. *Limit switches*, as these are called, work in the following way. Suppose that we want a robot to move its arm six inches to the left and four inches vertically and then come to rest. First, we will set a switch that will cause the arm to begin moving to the left as soon as the robot's power is turned on. Next, we will position a second switch exactly six inches to the left of the robot's starting point. The second switch will be activated by the robot's arm, which will come in contact with it when the robot's arm has closed the six-inch gap. The second switch will be wired to do two things when it is activated. First, it will shut off the power to the motor that moves the arm to the left. Second, it will apply power to another motor that moves the arm vertically. A third switch, set four inches above the second, will turn the robot off when the robot's arm comes in contact with it.

Robots that use limit switches and mechanical stops have point-to-point control, and are called *limited sequence robots*. Their movement between points, which is abrupt, has given rise to still another name. They are called, somewhat irreverently, *bang-bang machines*.

More sophisticated robots such as T³ are controlled by microcomputers. A computer is not a machine, such as a drill press, that is designed to do a particular thing. A computer is designed to have a certain capacity, and the computer's capacity allows it to be used in many ways—to play tic-tac-toe, for example, or to solve equations, or to prepare a budget forecast. Thus, it is not possible to say exactly what a computer can do. This also applies to computers that are used to control robots. Rather, it is possible to provide only illustrations of the ways in which such computers are being used.

For one thing, it is possible to replace devices such as limit switches with computer programs that control the robot in the performance of an entire task. Typically, these programs are stored on tape cassettes or disks. Changing the robot's task, as is done, for instance, when a factory has filled one order and begins to manufacture a different product for another order, can be as simple as replacing one cassette with another.

In addition, the computer can be programmed to vary the robot's routine as required by the task. For example, a robot may be programmed to perform quality control checks on the finished parts it removes from a machine tool and to place these parts on a conveyor, but to discard the parts it finds to be defective. Moreover, the robot may be programmed to call a technician when a defective part is found, or, perhaps, when it finds more than two defective parts out of every 100 parts produced. Typically, the robot will communicate with factory personnel by switching on a light in a control panel and, perhaps, by displaying a message that indicates the nature of the problem on the computer's cathode ray tube. Alternatively, the robot might be programmed to shut down the production process when defective parts are found, or to tell the technician about the problem by using computer-simulated speech.

In other applications, the computer may control several machines. With this type of control, each machine's operation can be coordinated with the operation of the other machines. In the preceding example, the robot will create a slight delay in its task cycle each time it recognizes and discards a defective part. The computer can be programmed to introduce this delay into every other machine under its control, so that all of the machines continue to be synchronized with one another.

Depending on the type of control system employed, it may also be possible to use the computer to keep track of defective work and to report scrap rates to management; to record the quantity of raw materials used in production, maintain a running balance of raw materials inventories, and reorder these materials whenever inventories reach redetermined minimum levels; and to maintain a continuously updated inventory of work in process.

At the extreme, an entire factory may be placed under the control of a large computer, which is linked to each of the micro computers that control the robots and other automated machines. One such plant is the Yamazaki Machinery Works in Nagoya, Japan. This factory requires few human workers. On the night shifts, the plant is occupied by a single watchman, who makes his rounds as the factory operates on its own making component parts for machine tools.

Other uses of computer control systems are equally impressive but less dramatic. For example, T³'s control system enables simulation of continuous path operation in a machine that moves from point to point in straight lines. This is accomplished by dividing the desired path between points A and B into numerous small straight lines that together approximate the path that one would have the robot follow as it moves from A to B. The greater the number of straight lines, or *chords*, between A and B, the more nearly the motion of the robot will approximate the desired path. In theory, this approximation technique could be employed by any type of controller, even one that consists of limit switches. In practice, though, only a computer can efficiently handle the large number of steps required by the technique.

Intriguing as they may be, though, computer control systems are not the *sine qua non* of robotics. Many tasks to which robots can be assigned call for nothing more sophisticated than the technology embodied in the simple bang-bang machines. As in other things, the degree of sophistication required of the robot's controller will be determined by the robot's task.

POWER

The robot's task also determines the kind of power used. Robots typically are powered by electricity, compressed air, hydraulic pressure, or some combination of these. In general, hydraulic systems are used in large robots that are required to perform accurately. Electric motors and similar devices are used in small robots. Bang-bang machines often run on compressed air. There are tradeoffs. Hydraulic power may be used instead of electricity when the robot must work in an explosive atmosphere and when there is a danger from electrical sparks. Electric power may be used in place of hydraulic power when there is a danger that leaking hydraulic fluid will contaminate the work in process. Pneumatic power is cheaper than either of the other forms of power and may be preferred for this reason. Once again, the task will determine the robot's specifications.

PUTTING A ROBOT TO WORK

Teaching a robot to work can be much simpler than training a human apprentice to perform the same task. In the case of continuous path machines, all that is required is to switch the robot to the teach mode and lead it by the hand through one complete task cycle. When switched back to the operating mode, the robot will continuously perform the task it has learned.

What this system of instruction implies is that robots can be taught skills that take years for humans to learn. For example, an experienced, highly skilled welder can teach a continuous path robot to weld exactly as he or she does. Every motion made by the welder while instructing the robot will be copied by the robot in subsequent performances of the task. Of course, the robot will mimic its teacher's faults as well. If the teacher's hand is unsteady after a hard night on the town, the robot's end effector will have the shakes until the machine is reprogrammed.

As the robot is led by the hand through a teaching cycle, the robot's controller monitors the position of the end effector, usually 60 times per second—the frequency of 60-cycle electrical current. Every robot has a home position—an end effector position that it remembers permanently—and the measurement of every position that the controller records is made relative to the home position. These recorded positions, in turn, are converted automatically to a program that the controller will follow in subsequent performances of the task, and the program is stored—usually on a tape cassette or disk. When the robot is assigned a new task, the disk or cassette can be retained for use when the robot returns to the original task.

Point-to-point robots are taught differently. The end effector is moved from one point in the task sequence to the next, using a hand-held control box that is provided for this purpose. Each of the buttons on the control box controls motion in one direction for one of the robot's joints. Contrary to the example in the previous paragraph, the controller records the location of only the points at which the end effector comes to rest, and it extrapolates a straight line between each pair of points—which is the path the robot will follow as it performs the task.

The method of instructing limited sequence robots was mentioned earlier. In addition, there are still other ways in which robots can be instructed, and occasionally several alternative modes of instruction can be used with a single robot.

The point is that although many robots employ state-of-the-art technology, they do not require great technological sophistication of the user. Similar to word processing machines and other forms of office automation that are designed to be "user friendly," modern robot design proceeds with the end user in mind.

The next chapter looks at the end user to provide an overview of the actual ways in which robots are used in industry. The uses described range from the replacement of a single worker with a robot to the integration of robots and other machines in unmanned factories.

3
Robots at Work

HOW A ROBOT PERFORMS A TASK

One of the enduring paradoxes in our economy is that jobs go unfilled even in times of high unemployment. Some of these jobs cannot be filled because they require occupational skills that are in short supply. The paradox is that other jobs, suitable for unskilled laborers, also go unfilled.

Unemployment rates typically are highest for those unfortunate workers who have the least education and work experience and the fewest occupational skills—who are, in other words, potential candidates for the jobs in question. What sort of job is so onerous that unskilled workers who are out of work will refuse to take it? One such job is that of operating an injection molding machine.

Plastic Molding: A Simple Application

Small, coin-operated plastic molding machines can be found in zoos, museums, and amusement parks. For 50¢, they will mold a plastic lion or giraffe before your very eyes. The way they work is as follows: A plastic tube that has been heated until it has become soft is fed between the halves of a mold of the animal to be formed. Next, the mold halves are forced together. As they join, they pinch one end of the plastic tube, sealing it. The other end of the tube, which protrudes through a hole formed by matching grooves in each half of the mold, is connected to a source of compressed air. When compressed air is forced into the heated tube, it blows up like a balloon. Thus, the tube expands until it fills the mold. As it cools, the plastic becomes rigid and maintains its molded form.

When the plastic has had sufficient time to cool, the mold halves separate from the plastic figure. The process is complete at this point, except that the figure still is attached to the end of the plastic tube from which it was inflated. A blade slices off the remaining tube, which is called a *sprue*, and nudges the molded figure into a chute from which it can be removed by the customer. Often, the finished figure will have a paper-thin, irregular ridge of plastic protruding from it. *Flash*, as this ridge is called, is formed when part of the plastic in the mold inadvertently is forced between the halves of the mold. It is removed easily with a pen knife.

Blow molding machines, akin to the one described, are used in industry to manufacture such products as plastic milk bottles. Solid plastic products such as eyeglass frames, pocket combs, and wastepaper baskets are made by the injection molding process, which is similar in many respects to blow molding. Injection molding differs in that molten plastic is forced into the mold, which it fills completely.

What, then, makes the job of operating an injection molding machine so onerous? For one thing, the work can be extremely uncomfortable. Unlike the coin-operated machine that is encased in a cabinet, industrial injection molding machines cannot be enclosed. The operator must be able to perform work between the dies, or halves, of the mold. Operating temperatures of such machines typically range between 200° and 300°C (390° to 570°F). In addition to being hot, the work area may be contaminated by noxious fumes given off by the molten plastic.

The work can be dangerous as well. Although some machines can be designed to eject the molded parts automatically, others cannot. In many cases, the operator must reach between the dies to lift out the molded part. Despite the presence of safety devices that are intended to prevent accidental closing of the dies—which typically operate at pressures of 14,000 pounds per square inch—the operator is exposed to the possibility of an accidental die closing each time he or she moves into the space between the dies. Operators of small injection molding machines thus risk the loss of hands or fingers as they retrieve small molded parts from the die area. Operators of larger machines—for example, those that mold plastic garbage pails—must place their bodies between the dies. Accidental die closings are rare, but they can occur.

Operating the larger molding machines can be physically taxing. Removing large objects from the die is hard work, especially when it is done in a hot, noisy, noxious environment. On top of this, the job is as monotonous as a job on an assembly line. A complete machine cycle, from one closing of the die to the next, generally takes less than a minute. In this time, the operator will open the dies, remove the finished part and close the dies, separate the molded parts if several are molded at the same time, trim the sprue and flash from each part, and dispose of each part by placing it on a conveyor or pallet or, perhaps, in a container. This series of tasks is repeated over and over on each shift—for the larger, slower machines, 60 or more times an hour, 480 times a day, and 2,400 times a week from week to week and year to year. Plastic molding machines, as I indicated, operate at relatively high temperatures. Because it takes

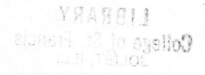

several hours to heat a cold machine to the correct operating temperature—hours during which production is lost, since parts molded at lower temperatures often are defective—it is common to operate these machines continuously across three shifts.

A Robot to the Rescue

How do organizations recruit reliable workers to fill jobs that are dull, repetitive, unpleasant, potentially dangerous, and that provide relatively modest wages? As you might suspect, such positions are difficult to fill.

Imagine, then, the enthusiasm with which a plant personnel director might receive the following resumé vita (Figure 3–1) from a job applicant.

Unimate 2000B is something of an aristocrat in the world of robotics. Unimation, its builder, is the oldest and, presently, the world's largest supplier of industrial robots. The first Unimate was sold to General Motors in 1961.

As depicted in Figure 3–2, 2000B is designated to provide six degrees of freedom. Looking more like the gun turret on a Sherman tank than a mechanical man, this robot, time and again, has demonstrated its utility as an economical substitute for human labor.

One of the early substitutions of robots for human labor was in metal die-casting operations, where robots have logged over a million hours of production.[1] More recently, robots have been used to operate injection molding machines, which are quite similar to die-casting machines.

Major differences in the ways in which robots are used to operate injection molding machines occur because of dissimilarities among molding processes. For example, the time between die closing and opening in a small molding machine may be so brief that a robot will have barely enough time to remove the part from the die and dispose of it before the next part has been made. Larger machines, with longer cycle times, will permit the robot to carry out many more tasks. In order to demonstrate how robots work, I will describe an operation that is hypothetical but nonetheless very much in line with actual practice.

[1]Goldhamer, W.M., "A Systems Approach to Robot Use in Die Casting," In W.R. Tanner (Ed.), *Industrial Robots Vol.2 Applications*. Dearborn, MI.: Society of Manufacturing Engineers, 1979, pp. 69–72.

<u>VITA</u>

UNIMATE 2000B

Shelter Rock Lane
Danbury, CT 06810

Social Security No.: None
Age: 300 hours
Sex: None
Height: 5 ft. Weight: 2800 lbs.
Life Expectancy: 40,000 working hrs.
                  (20 Man-shift yrs.)

<u>Position Desired</u>: Die Cast Machine Operator

<u>Salary Required</u>: $4.00/hr.

<u>Other Positions for Which Qualified</u>:
  Forging press, plastic molding, spot welding, arc welding, palletizing, machine loading, conveyor transfer, paint spraying,
  investment casting, heat treatment, etc.

<u>Education</u>: On the job training to journeyman skill level for all
           jobs listed above.

<u>Languages</u>: Record-playback, Fortran, assembly

<u>Special Qualifications</u>: Strong(100 lb. load), untiring 24 hours per
                          day, learn fast, never forget except on com-
                          mand, no wage increase demanded, accurate to
                          0.05" throughout range of motion, equable,
                          despite abuse.

<u>History of Accidents or Serious Illness</u>:
  Suffered from Parkinson's Disease(since corrected), lost hand(since
  replaced), lost memory(restored by cassette), hemorrhaged(sutured
  and fluid replaced).

<u>Physical Limitations</u>: Deaf, dumb, blind, no tactile sense, one armed,
                        immobile.

<u>Notify in Emergency</u>: Service Manager, Unimation Inc., (203) 744-1800

<u>Dependents</u>: Human employees of Unimation, Inc.

<u>References</u>: General Motors, Ford, Caterpillar, Bobcock Wilcox, Xerox,
              and 65 other major manufacturers.

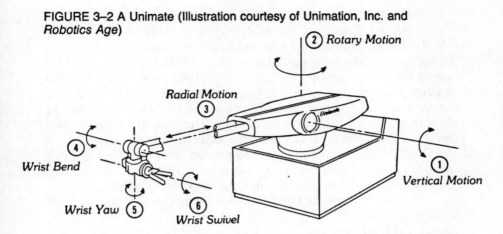

FIGURE 3–2 A Unimate (Illustration courtesy of Unimation, Inc. and *Robotics Age*)

Six program controlled articulations of Unimate industrial robot.

A JOB AT WHICH ROBOTS WORK

The operation in mind produces cases for hand calculators that are about the size of a deck of playing cards. Each case consists of a front and back half, which snap together. Both halves of the case are molded simultaneously. Molten plastic is forced through the sprue into a channel called the runner.[2] *The runner carries the plastic into the dies that form each half of the case. The molded part that emerges from the dies consists of the two halves of the case and a connecting solid rod of plastic, the runner.* The sprue, another bar of plastic, is connected to the mid-point of the runner at right angles to it. You probably have seen a plastic model airplane kit. The model's parts, which normally come in sheets, are connected by runners.

The robot that operates this injection molding machine begins a cycle of work by signaling the machine to open the dies. It does this by sending an electrical impulse to a switch that controls movement of the dies. As the dies open, ejector pins automatically raise the molded part until it is clear of the lower half of the die. These pins are built into the injection molding machine.

---

[2]Both the channel and the plastic that remains in it when the molten plastic cools are called runners. The same is true of sprues.

As the dies open, the robot begins to move its end effector into the die area. When the dies are fully separated, the robot grips the molded part by its sprue. The sprue provides a convenient place to grip the part, since it will not matter if it is marred by the end effector. The position of successive molded parts normally will be the same no matter how many parts are made. This is important because our robot cannot see the part it grasps. It cannot notice changes in the positions of parts and adjust to these changes.[3]

It is important to be able to predict with accuracy the orientations of the objects that are to be processed by robots. Die casting and injection molding provide "natural" tasks for robots because the orientations of the molded parts, as they lie on the ejector pins, tend to be invariable.

Nonetheless, one must acknowledge Murphy's Law. Suppose, for example, that a molded part fails to eject from the die, or that the sprue breaks and remains stuck in the die. The robot must be equipped to sense whether or not it actually has lifted a part from the die. If the robot is not equipped to do this, and, for example, a sprue remains in the die, the robot will continue to work as if nothing out of the ordinary had happened, with unfortunate consequences. In our example, the robot could be equipped to sense overgrasp in its end effector. The end effector of our robot is similar to a pair of tongs. As the tongs grasp a sprue, they close to a certain point. Should they continue to close beyond this point, it would be safe to assume that, for one reason or another, they failed to make contact with a sprue. A robot that is equipped to sense overgrasp will generate an electrical signal if its tongs close beyond the predetermined point, and this signal will set the following sequence of events in motion: the robot will (1) withdraw its arm from the dies, (2) turn the injection molding machine off so that the dies do not close again, and (3) signal an operator that a malfunction has occurred.

There are numerous alternative ways to determine whether a part actually has been grasped by the robot. Some of these enable the robot to kill two birds with one stone, so to speak. Rather than equip the robot in this example with the ability to sense overgrasp, we will consider one of the alternatives.

Once the robot has grasped the sprue, it will remove the

[3]Such adjustments can be made by robots equipped with vision systems, which are discussed in Chapter Five.

molded part from the die area. At this point, Murphy's Law must be confronted once again, for it is entirely possible that the sprue and only one half of the case have been lifted out of the die area while the other half, having broken off, has remained in the die. Alternatively, the robot may have come away from the die empty handed. To sense whether or not either possibility has occurred, our robot will move the molded part past two feeler switches. One of these switches will come in contact with each half of the completed part. If either half of the part is missing, or if the entire part has been left in the die area, one or both of the switches will fail to make contact. In either event, a signal will be generated to set into motion the steps enumerated.

When the robot knows that it has removed a complete part from the die, it will also know that its end effector is clear of the die area. Thus, it will know that it is safe to close the dies, which it will proceed to do. As the dies close and plastic is injected between them, the robot will place the part it holds in a trim press. This press is designed to slice the flash from each half of the casing and to separate the casing from the runners. As the robot's arm moves away from the trim press to a safe position, the robot will signal the press to close. The molded parts, divested of runner, flash, and sprue, will drop onto a conveyor belt that will carry them away from the robot's work station. The runner, sprue, and flash will fall from the press into a scrap bin. As this occurs, the robot will return its arm to the vicinity of the injection molding machine and wait for the proper moment to open the dies.

Exploiting the Robot Worker

Depending on the cycle time involved, the robot may be left with a little time on its end effector at this point. If the interval of time between operating the trim press and opening the die is long enough, it may be possible to have the robot operate a second injection molding machine. The steps followed in operating the second machine are similar to those described earlier. In addition, the robot will be programmed to alter its behavior in the event that a problem occurs in one or both of the injection molding machines. This can be accomplished by elaborating the branch in the robot's program that has been described. The original branch caused the robot to stop and signal an operator in the event of a machine malfunction. This branch will be replaced by a three-way branch. As before, each part will be tested as it is removed from the dies. In the event that a part from the left-hand machine is found to be defective, the robot will turn off that machine and signal an operator while it continues to operate the right-hand machine. If a part from the right-hand machine is found to be defective, the robot will stop that machine, call the operator, and continue to operate the other one. Should both machines malfunction, the robot will stop everything and call for help.

Variations on a Robot's Job

At this point, I will mention some of the major variations in injection molding processes in order to demonstrate the robot's flexibility. Such variations could easily occur in the application described in the foregoing section if the manufacturer produced plastic parts in batches for a number of customers having diverse products.

When the dies are changed prior to the start of a new production run, a number of corresponding adaptations must be made to the robot's program. For one thing, the cycle time for the injection molding machine may change if the new parts are significantly larger or smaller than the parts previously molded. For another, the size and shape of the sprue may vary from one set of dies to the next. Moreover, it may be necessary to have the robot handle the molded part by one of its runners rather than by its sprue. Obviously, such changes can be accommodated by changing the robot's program, although it may be necessary to replace its end effector as well if the present end effector cannot adequately grasp the new part.

Many injection molded parts contain inserts. For example, a plastic wagon wheel may include a metal bearing; eyeglass frames often are molded around wires that are incorporated to add strength to the plastic part; and so on. Such inserts are placed in the dies, where they become embedded in the molten plastic. Our robot can be programmed to do this as well. However, unlike a human worker, the robot will not be able to remove inserts from a bin or box and place them in the die. Rather, the robot must know the location and position of each insert it grasps. One way to handle this problem is to deliver the inserts to the robot on a conveyor that funnels each insert to a predetermined pick-up point. Another approach to the problem is to deliver batches of inserts to the robot on pallets. In this case, each insert on a pallet will have a different location with respect to the robot, but each location will be predetermined. The robot will be programmed to grasp successive inserts in successive locations. This is more or less like placing the inserts on a chess board and telling the robot to grasp the first insert from the white queen's rook's square, the second from the white queen's bishop's square, and so on until the board is empty. When the board has been emptied, the robot will assume that the board has been replaced by a new one and will grasp the next insert from the white queen's rook's square. Throughout this operation, the robot will act as if each insert is in the correct position, which it must be if the operation is to succeed. Thus, each insert, regardless of whether it is on a pallet or conveyor, must be right side up and facing the proper direction.

Before the robot places an insert in the die, it must determine whether it actually has hold of one. As was the case before, overgrasp sensors, feeler switches, or similar devices can be used to provide this information. In addition, the robot's program will now include another branch that will cause the robot to take appropriate steps if it finds that it does not have an insert in hand when it needs one.

As mentioned earlier, there is no need to build robots that look like people. Thus, if it will be advantageous to have a robot with two hands on its arm, there is no reason not to build one with two hands. In fact, two-handed robots sometimes are used to save time in injection molding applications. One hand removes the part from the die and the other places an insert in the now empty die. A one-handed robot would be required to remove the part, dispose of it, grasp an insert, and return to the die to place the insert in position. Since the robot must reach into the die area twice during each cycle, the cycle time would be increased. Use of a second hand can eliminate the need to delay the die closing and, thereby, allow the injection molding machine to maintain a higher level of output than would be possible otherwise. In other cases, the end effector that removes the part from the dies may not be able to hold the insert. The need for two hands in such cases is obvious.

On removing a part from the machine, the robot may perform a series of quality control checks. The feeler switches mentioned provide, among other things, a check on the completeness of a molding. Other similar checks can be made, and the robot can be programmed to deliver defective parts to a scrap bin rather than to the trim press, for example.

Finally, the robot may be programmed to palletize the manufactured parts or to place them in cartons. In either case, the trim press would be designed to hold the finished parts rather than to drop them onto a conveyor. The robot would remove each half of the casing from the press and place it on a pallet or into a box. Problems and solutions similar to those encountered in grasping an insert will be encountered in palletizing or packaging finished parts.

Functions of a Robot

Before discussing other types of jobs to which robots have been assigned, it will be helpful to review the tasks described in the previous illustration to place them in a much broader context. When one steps back from the subject, robots appear less as machines than they do as surrogates for human workers.

Unlike other machines that follow blindly a sequence of predetermined steps, the robot in the preceding illustration is able to modify its behavior to respond appropriately to changes in its external environment. For example, the robot is able to perform quality control checks on the output of the injection molding machine and to alter its performance in the event that a defective part is produced. The ability to sense and react to the external environment can be considered as a rudimentary example of machine intelligence.

Next, the robot performs its task by, among other things, operating other machines: the molding machine and trim press. Moreover, it does this while following safety rules; for example, it closes the dies only after it has observed that its end effector has passed the feeler switches and is clear of the die area.

Third, the robot is able to communicate with plant personnel. It can bring problems to their attention and even describe the nature of the problem at hand.

Finally, the robot can learn new behaviors and, in effect, can remember old behaviors, as is the case when its program cassettes are changed. To appreciate the significance of this last point, one must make a distinction between what is feasible using present technology and what is economically justifiable. If we replaced the robot's microcomputer with a larger one, it would be possible to store all of the robot's programs in the computer. In this case, the robot would have the ability to recall every task it had ever learned. All that would be required of the operator would be to change the dies and to tell the robot about the change. Assuming that it previously had manufactured the part in question, the robot would look up the correct program, perhaps ask the operator for a different end effector, and go to work.

Here, then, is a machine that has the potential to learn and remember, to adapt to its environment, to use other machines, and to communicate with people. It is not just a machine: it is a new technology—one that may reshape the way in which we live and work. Such thoughts easily shade off into hyperbole. Yet, the field of robotics stands today where the computer industry stood in the 1950s. The development of robot technology and the growth of this industry are expected to parallel the growth experienced in the plastics industry in the 1950s, the automobile industry in the 1960s, and the computer and microcomputer industries in the 1970s.

I worked with vacuum tube computers in the early 1960s. Even then, just a decade after they were manufactured, they were anachronisms. These machines barely hinted at the evolution of computing equipment that was to follow. They performed simple arithmetic functions and punched their meager results into cards. Today, you can buy a more powerful calculator with a $10 bill and get some change back.

Nonetheless, the early vacuum tube machines opened new avenues of research for a handful of scientists who used them in work that led to the development of subsequent generations of computers. Vacuum tube computers were even used in early research on artificial intelligence. The development of computer technology throughout the subsequent 20 years was nothing short of astounding. Similar developments are predicted for the robotics industry. T³ and Unimate 1000 are only the beginning.

4

Economics and Robots

ECONOMIC AND TECHNICAL
ANTECEDENTS OF THE ROBOT INDUSTRY

Variations on an Ancient Theme

Most inventions have their origins in ancient fantasies and dreams. Similar to the idea of human flight, which was dramatized in the Greek myth of Icarus and probably first took hold in the dawn of civilization, the notion of a robot, an artificial being, has been with us since time out of mind. Pandora, for example, as well as Golem and Dr. Frankenstein's monster were depicted as artificial creations rather than as natural beings. By the beginning of the present century, robots had entered the mass media. Between 1897 and 1919 at least 29 motion pictures in which artificial beings were portrayed had been produced.[1] The term *robot* was coined in 1920 by Karel Capek, author of the play *R.U.R. (Rossum's Universal Robots).*[2] Today it is difficult to avoid daily references to the robotics industry unless one chooses to forsake newspapers, journals, and the electronic media.

The present level of interest in robotics is difficult to justify, though, when one looks merely at the industry's current level of activity. As we know, the world robot population is approximately only 30,000. There are more than 30,000 bathroom scales in Topeka, Kansas.

Some 20 years after its beginning, the robotics industry still is struggling. World-wide, the number of robots sold in 1980 was less than 4500. Total sales revenues in that year were only $100 million. The industry is not exactly awash in profits. Why, then, all the fuss about robots?

An Idea Whose Time Has Come

The answer lies in an analysis of a number of trends that lead to expectations of extraordinary growth in this industry. The $100 million in sales experienced in 1980 became $155 million by 1981 and is expected to exceed $200 million by the end of 1982. However, recent forecasts project annual sales revenues, world-wide, of between $3.5 and $5 billion by 1992.

[1]Robert M. Hefley, *Robots.* New York: Starlog Press, 1979.

[2]*Robot* was coined from the Czechoslovakian word, *robota,* which means servitude or forced labor.

Such forecasts rest, as I indicated, on a number of independent trends that together create expectations of a changed economic environment in which this industry will flourish. The oldest of these trends began early in this century with the introduction of scientific management. One aim of scientific management was to reduce work to its component tasks to find efficient methods of performing them. The result is exemplified by the assembly line. An unintended consequence of scientific management was to simplify and routinize such tasks to the extent that their successful performance no longer required many human aptitudes. They become simple enough that it looked as if machines might be able to perform them.

Be that as it may, what is possible is not always profitable, and for many years human labor had an economic advantage over machines. Such advantages do not last forever, however. Within the past 10 years, labor costs in this country have increased by approximately 250 percent, while the costs of manufacturing robots have increased only 50 percent. Today there are tasks for which robot labor clearly is more economical to employ than human labor. Moreover, this trend is predicted to continue and to accelerate.

Here, the trend lines I mentioned earlier form a warp, as it were, for the fabric of society in the 1990s. First, there has been an inexorable upward pressure on salaries and wages stemming from both inflation and the ever-increasing aspirations of the work force. At the same time, labor productivity in the United States has declined, perhaps as a result of a gradual erosion of individual initiative but surely as a consequence of the steady decline in our rate of capital investment. The expense of human labor has been increased further in the past several years by increased costs of fringe benefits and by the costs of complying with federally mandated regulations such as occupational health and safety standards.

The Robot's Competitive Advantages

Relative to human workers, robots have become economical. No doubt economies of scale have been realized in the early growth period of this industry and have helped to hold down costs. More important in this regard, I think, have been the gradual improvements in robot technology. By this I mean the technologies associated with designing and deploying robots. Manufacturing technology in this industry is, if anything, behind the times. The Fujitsu Franuc facility in Japan, in which robots labor around the clock building other robots, is a unique exception. In contrast, until 1981 the United States's oldest and largest manufacturer of robots, Unimation, ". . . was organized as a large job shop. Each robot was built as if it were custom designed from top to bottom." [3]

Most notable in the realm of robot technology has been the growing use of microcomputers as control devices. Machines that use mechanical control equipment such as limit switches (the so-called bang-bang machines) or fluid controllers are rapidly becoming anachronisms. Computer control obviously is superior in terms of capacity, speed, reliability, and a number of other important criteria. Until fairly recently, though, it was prohibitively expensive. Obviously, that is no longer the case. The drastic reduction in the cost of computing equipment, which has been nothing short of astounding, is summarized by MIT's Edward Fredkin as follows: "The trends are very clear—they'll continue to the end of the century, close to a factor of two every year. Which means every ten years a factor of a thousand, every twenty years a factor of a million. Today's quarter-million-dollar system will cost $250 in ten years, and will be better besides." [4]

Computer control of robots has greatly simplified the task of programming them to perform their tasks. In certain applications, such as welding and spray painting, it is possible to teach a robot in minutes skills that human laborers require years to learn. Moreover, having learned to perform a task such as spray painting, a robot often is able to out-perform its human counterpart. The reason is that robots, unlike you or me, do not vary the way in which they work. If a robot is trained to perform a task by a skilled worker who strives to attain optimal performance at the task while instructing the robot, then, barring a malfunction, the robot will replicate within narrow limits the optimal performance of the task each time the task is executed. Human task behavior is less reliable and is subject to greater variance. Workers get bored, tired, or distracted, and their performance suffers as a consequence.

Within limits, a robot can function in environments that would be dangerous, or at least uncomfortable, for human workers. The major implications are that: (1) robots can perform tasks in environments more nearly suited to the requirements of the task (e.g.,

[3]John W. Dizard, "Giant Footsteps at Unimation's Back," *Fortune*. Vol. 105, No. 10, May 17, 1982, P.97.

[4]Edward Fredkin as quoted in Pamela McCorduck, *Machines Who Think*. San Francisco: W.H. Freeman and Co., 1979, p.349.

they can spray paint in higher ambient temperatures than can humans, thereby permitting better quality in the paint-spraying operation), and (2) the cost of maintaining a suitable environment for robots often is less than it would be to maintain an environment for humans employed at the same tasks. To continue the paint-spraying illustration, a robot will require less ventilation than will a human worker, since the robot does not breathe. A higher level of air contamination will be permissible. Less light may be needed at the work station, since the robot cannot see. Higher noise levels may be permitted. These and other relaxations of environmental standards can provide substantial savings for the firm that substitutes robots for human labor.

Now, add to the foregoing advantages the following: Robots can work around the clock and do not receive differential pay for night work; they are not prone to go on strike or to display other forms of labor unrest; they occasionally break down, but never get sick and are immune to the adverse effects of drugs and alcohol; and they are readily available to work at unpleasant tasks for which human laborers are difficult to recruit.

And Disadvantages

The point that robots break down should not be overlooked, for it constitutes one of the major drawbacks in substituting robots for human labor. If a robot malfunctions, and it will, one of two things will occur, depending on the nature of decisions that were made at the time the firm using the robot was planning for its installation. Either a substitute for the malfunctioning robot will take its place in the work process until repairs have been completed, or the robot's work, as well as that of human employees and machines whose functions are contingent on the malfunctioning robot's output, will grind to a halt until the robot goes back on line.

Early applications occasionally permitted human workers to take the place of disabled robots, but this rarely is the case today. For example, today's industrial robot may work in conjunction with other computer-controlled machines that are synchronized by their control systems. In such cases, it is not possible to back up these machines with human operators, and when one machine goes down, they all will go down.

Thus, the expected down time for robots and related equipment is of crucial importance to users. Down time is a function of the mean time between equipment failures and the mean time needed to repair equipment that has failed. Industrial robot applications are difficult to justify unless expected down time is 2 percent or less.

ECONOMIC COSTS AND BENEFITS
OF ROBOT WORKERS

The Robots' Niche

The robot's advantages over human labor are limited to classes of tasks that fall within the robot's range of abilities—tasks, for example, that do not require a well-developed vision system. Moreover, to compete effectively with human labor, a production run of some minimum size is required. The costs associated with reprogramming and retooling robots and associated equipment when production runs are changed make them prohibitively expensive to use in small-batch operations. The general relationships between volume of production and unit cost for human labor, robot labor, and conventional automation are depicted in Figure 4–1.

Figure 4–1 also depicts that fact that, in the case of lengthy production runs, conventional automation such as transfer assembly equipment is more efficient to use than are robots. Thus, it is in production runs of intermediate length that robots can be used to advantage.

FIGURE 4–1 Volume of Production vs. Unit Cost for Human Labor, Robot Labor, and Conventional Automation.

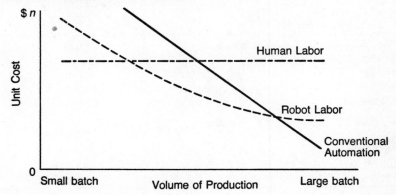

The relative efficiency of robots vis-à-vis human labor can be shown more clearly in the following illustration. Here, the spray-painting example is continued, and a comparison of the costs of human and robot labor demonstrates the savings that may be gained by employing the latter. Data used in this illustration are estimates based on cost analyses published elsewhere.

Costs

Robots range in price from $1,600 to upward of $600,000. A number of paint-spraying robots are available for $100,000 and less. A cost of $90,000, which may be somewhat on the high side, has been used in Table 4–1. Besides the $90,000 required to purchase the robot, its end effector (which in this case is a spray gun), the controller, and other accessories, an additional investment of $10,000 is required to adapt the plant for the robot's installation. Included in this estimate are the costs of the pad on which the robot will be mounted, electrical

TABLE 4–1. The Economics of Robot Labor: An Example

Capital Investment		One-Shift Operation	Two-Shift Operation
Cost of robot and accessories	(a)	$ 90,000	$ 90,000
Installation costs	(b)	10,000	10,000
Total: a + b =	(c)	$100,000	$100,000
ANNUAL COSTS			
Depreciation = c ÷ 7	(d)	$ 14,285	$ 14,285
Maintenance cost	(e)	3,500	7,000
Operating and programming costs	(f)	3,500	7,000
Insurance	(g)	2,000	2,000
Total: d + e + f + g =	(h)	$ 23,285	$ 30,285
ANNUAL SAVINGS			
Materials savings	(i)	$ 12,000	$ 24,000
Wage and fringe benefits at			
$10/hour	(j)	20,000	40,000
$15/hour	(j_2)	30,000	60,000
$20/hour	(j_3)	40,000	80,000

Total Annual Savings = $(i + j_n) - h$; ROI = $\dfrac{[(i + j_n) - h] \times 100}{c}$

at $10/hour	$ 8,175	ROI = 8.175	$33,175	ROI = 33.175
at $15/hour	18,175	18.175	53,175	53.175
at $20/hour	28,175	28.175	73,175	73.175

or other power connections, safety fixtures such as a protective railing around the robot, and the like.

Annual Operating Costs

The $100,000 investment is charged against current operations over seven years according to a straight-line depreciation schedule. Maintenance cost, which includes both labor and parts, is calculated at a rate of $1.75 per hour of operation. Operating and programming costs include the cost of energy that is needed to operate the robot as well as a rough estimate of the labor costs involved in programming and reprogramming the robot. Insurance is estimated to cost $2000 per year.

Annual Savings

A robot may perform its task more or less efficiently than the worker it replaces. In the case of paint spraying, robots typically outperform people. A considerable amount of paint is lost in spraying operations through what is called overspray: the spraying of paint past the object being coated. Robots can attain a higher degree of accuracy in their repetitive movements than people can, and, thus, overspraying can be reduced. The reduction in the cost of overspray of $12,000 per shift per year is based on an estimate published by the Society of Manufacturing Engineers.[5] There are still other benefits to be gained from substituting robots for people in this type of task. For instance, robots usually are able to apply a more uniform thickness of paint to the surface being coated. In many paint-spraying operations, the associated improvement in quality is of considerable value to the producer. In other cases, the robot's greater accuracy of movement will reduce the rate at which work is rejected due to inadequate finishing. The estimated cost saving used in this illustration is conservative in that it considers only a reduction in the cost of wasted paint, and does not account for the savings that result from improvements such as those noted above, the energy savings that result when overspraying and reject rates are improved, or the savings that can accrue when temperature, airflow, and illumination standards that were maintained for human workers are relaxed.

Annual labor savings are computed for cases in which wages

[5]Timothy Bublick, "The Justification of an Industrial Robot," *Industrial Robots, Vol.I, Fundamentals*. Dearborn, Mich.: Society of Manufacturing Engineers, 1979, pp.39–45.

and fringe benefits equal $10, $15, and $20 per hour. An eight-hour
shift and 250 work days per year are assumed.

Return in Investment

The approximations of return on investment in the foregoing illustration, although crude, suggest the ultimate reason for industry's
growing interest in robotics. For each of the two-shift applications
considered here as well as for single shift applications where labor
costs are in the neighborhood of $15 per hour or greater, the pretax
ROI estimates provide strong inducements to consider replacing human workers with robots, and that is what all the fuss is about.

Some Caveats and Rules of Thumb

Table 4–1 provides a much simpler analysis of return on investment
than actually would be used in evaluating the alternative of using a
robot. Moreover, while the data are fairly descriptive of some robot
applications, they cannot be used as the bases for broad generalizations. The range of variance in such data from one application to
another will help to make this point.

   The installation cost of $10,000 is appropriate in cases where
little modification of existing facilities is required by the robot. Yet,
where this has not been the case—where extensive factory modifi-
cations have preceded installation of a robot—such costs have run
as high as three times the robot's purchase price.

   To estimate depreciation costs, I have assumed that the hypothetical robot will be operated well below its designed load limits,
and that, consequently, it can be used for seven years without over-
haul. Many of the robots in use today are depreciated over ten years,
but are overhauled at the end of five years of two shift operation.
Such overhauls can cost twenty-five percent as much as the robot.
Major overhauls may be needed sooner for robots that operate close
to their design limits. One rule of thumb suggests that the need will
occur after 15,000 hours of operation in such cases, where 40,000
hours is taken to be the robot's useful life.

   There is also a cost to learning to use this technology, which
is not contained in the previous table. In many cases, management
must learn to solve unique problems before a robot can be used
effectively. In the case of paint spraying, for example, a problem

regarding the lack of uniformity in raw materials may be encountered. The viscosity of paint will vary over time—sometimes from
one shipment of paint to the next and at other times as the ambient
temperature and humidity in the plant vary according to the season.
Workers are able to detect such changes in raw materials and to
adjust the way in which they apply the paint to accommodate them.
But a robot is unable to see the finish that it applies and, consequently, is unable to change its work methods accordingly. Therefore, it must be supplied with paint of unvarying viscosity. Thus,
different types of paint may be needed to achieve the same finish at
different times of year. In addition, it may be necessary to convince
paint manufacturers to supply a product of more uniform quality.
Such lessons are learned from experience and are learned at some
cost. Yet, the problems from which they arise vary so greatly from
one application to the next that it would be pointless to try to generalize about their costs to management.

 While I am cataloging my sins of omission, let me add two
more. The operating costs used in Table 1–1 are not an estimate from
which generalizations can be made. What one can say in general
about such costs is that they are based, in part, on retooling and
reprogramming costs that, in turn, are determined by the amount of
time that these activities consume. In general, the time involved
should not exceed ten percent of the length of the production run.

 Finally, I have used a constant hourly cost of labor in Table
1–1. In practice, one would use an average expected labor cost for
the depreciation period. To this, one would add the amortized costs
of staffing the position in question, which would include recruiting
and training expenses as a function of labor turnover.

 Such caveats notwithstanding, some 30,000 robots are used
today—the great majority to the economic advantage of the firms
employing them. More will be employed with each day that passes
until the nature of work is transformed.

FUTURE ECONOMIES FROM ROBOTICS

A New Industrial Revolution

We seem to be on the verge of another industrial revolution. According to Dr. Frank Press, President of The National Academy of
Sciences, "... the next twenty years will witness a new industrial

revolution based on high technology that will prove to be as important as those preceding it." [6] *Fortune* stated it more dramatically by claiming that the technology which is characterized by computer controlled robots ". . . has more potential to increase productivity than any other development since electricity." [7]

Whether the United States will benefit as much as other industrial nations from the coming revolution will depend to a very great extent on events that will be outside of the control of this nation's businesses. Japan presently is the world's leader in building and using robots. Sixty-nine percent of the approximately 30,000 robots in use today work for the Japanese. Ironically, Japan bought her robot technology from a United States firm, Unimation. High interest rates, uncertainty concerning future tax regulations, and other government controlled or influenced factors inhibit American businesses from making capital investments in equipment such as robots. The Japanese government, on the other hand, through its Ministry of International Trade and Industry, provides low interest loans to small and medium sized industries to encourage their use of robots and related technologies.

Japan, as well as Germany, France, and Russia, are training workers who will enjoy greater technological literacy than will the average American worker. According to the American Academy of Sciences, Japan graduates twice as many engineers each year as we do. But our population is twice that of Japan. Forty-five percent of Russia's university graduates major in science, mathematics, or engineering, but only 15 percent of ours do. Half of the precollege teachers of science and mathematics newly employed in this country, according to a 1981 study, are formally unqualified to teach those subjects. The high school my son attends did not offer a course in physics until 1982, and yet, the ability to develop and use high technology rests on the availability of educated scientists and technicians.

We now have unemployment in the neighborhood of 10 million people. One can anticipate, in light of this, social opposition to the wide-spread introduction of a technology designed to put even more people out of work. Yet, in opposing implementation of the technology, the nation could find itself less able than ever before to compete with other industrial nations, and worse off, due to a re-

[6]Frank Press, adapted from remarks to the Washington Press Club, July 20, 1982.

[7]G. Bylinski, "A New Industrial Revolution is on the Way," *Fortune*. Vol. 104, No. 7, Oct. 5, 1981, P.114.

sulting slow growth in the economy, in terms of its ability to provide jobs.

The robot is only one member of a family of machines that includes equipment used in computing, computer-controlled manufacturing, and computer-assisted design. The ultimate use of a robot, I think, is not merely to replace a worker in an otherwise unchanged factory but to take its place in a fully automated factory, such as exist in a few places today, to relieve us from dull, repetitive labors and to increase our ability to produce.

A Future for Robotics in America

The benefits of robotics are not so much at hand as they are in our future. What the future will be is anyone's guess, according to some people, but I believe otherwise. As a behavioral scientist, I have come to understand much human behavior to be the product of unconscious and, in in a sense, irrational mental processes. And yet, certain kinds of behavior clearly result from rational thought or at least from the intention to behave rationally. These behaviors are typified by, but are by no means limited to, certain activities that people are taught to perform. Many aspects of work in accounting, engineering, or management, for example, can be understood as rational attempts to develop optimal or at least workable solutions to well-defined problems.

It should be possible, then, to construct a fairly accurate scenario of the future of an industry by projecting the future consequences of the logical, rational responses that the industry has at its disposal to the problems and opportunities confronting it today. In order to do this, though, it first will be necessary to make some major assumptions about the future of social processes that do not appear to be quite as rational as economic decision making in a firm. Namely, one must make some assumptions about the problems that are outside of the firm's sphere of direct influence. One must assume, for example, that the quantity of capital available for future business investments will be sufficient both to replace existing high-interest debt and to support direct capital investments in technology. Second, one must assume that corporate income tax regulations will be stabilized to the extent that managers will be able to make reasonably accurate estimates of after-tax returns on such investments. Third, it is necessary to make the heroic assumption that current trends in public education will undergo discontinuous changes, that either public educational institutions or private employers will redress the present shortcomings of our educational system. I am fairly sanguine that these and certain other assumptions that must be made will prove to be more valid than not. This country's shortcomings are becoming evident to all, as is the certainty that in failing to redress them, we will find our standing among other nations gradually but continually eroded.

Having made the kinds of assumptions that must be made if continued economic growth is to be predicted for the nation as a whole, I will turn to the problems and opportunities confronting industry and examine the means that presently are available to solve or exploit them as the case may be. The purpose in doing so at this point is to examine some likely economic benefits that robotics may provide. Two related aspects of the future are examined in subsequent chapters. In Chapter 5, present research and likely improvements in robot technology are discussed. The social consequences of widespread adoption of this technology are explored in Chapter 7.

Trends in Robot Employment

Till now, the majority of robots employed in this country have been used in two types of applications—spot welding and materials handling.[8] Spray painting and assembly applications are just about tied for third place as users of robots. However, according to *Robotics Age*, the majority of robots purchased by 1990 will be assigned to materials handling and assembly work.[9] The implication, if one believes current industry forecasts, is that the present market for assembly and materials handling robots of approximately $80 million per year in sales will grow, at the outside, to $3.7 billion per year by 1992.[10]

[8]Under the category of materials handling applications are included various machine loading and unloading operations.

[9]Laura Conigliaro, "Bullish Days in the Robot Business," *Robotics Age*. Vol. 3, No. 5, Sept.–Oct. 1981, pp. 4–7.

[10]A brief note on large numbers. If you were given $1000. a day and told to spend it all, you would run through $80 million in about 250 years. Two hundred and fifty years ago, in 1732, Plymouth Colony was celebrating the 102nd anniversary of its founding. However, spending $1000. a day, it would take you just over 11,000 years to unload $3.7 billion. Eleven thousand years ago, in 9000 BC, the Neolithic period had begun: farming was being invented as were refinements in stone tool making. Neither the Great Wall of China nor the Pyramids had been built. Indeed, over seventy centuries would pass before the pyramid of Tutankhamen would appear. As this analogy suggests, 3.7 billion is an immense number relative to 80 million.

How does one justify a forecast for such an enormous rate of growth? For an answer, one must look to fundamental changes that are occurring in manufacturing—changes that began in the early 1960s.

In 1960, I earned my baccalaureate degree in marine engineering and took a berth on a 12,000 ton freighter. As an engineering officer, I operated and maintained a steam plant that exemplified nineteenth century technology, and as I labored in the engine room amid reciprocating steam pumps and boilers, the age of space travel was dawning and the computer age was upon us.

About that time, manufacturing engineers were envisioning ways in which computers could aid them in their work. By the middle of the decade, the computer was seen as a potentially significant aid in drafting, which is one of the tedious, time-consuming tasks in engineering. The notion was that a draftsman might sketch a rough design on a cathode ray tube attached to a computer. The computer, in turn, would redraw the line precisely, according to a predetermined program and previously entered data that described the part being drawn. In this manner, the manhours spent at the drafting table could be reduced, particularly in cases where design changes required one modification of drawings after another.

This use of computers, although ingenious, would not have been particularly notable except for the fact that it contained the germ of another technology. The data used by the computer to draw blueprints could also be used to design and test the part in question. Moreover, these data could be used, ultimately, to control the machines used in manufacturing the part.

Engineering design can be an extremely complex task, as the following simple illustration will show. Imagine, for example, the basic problem involved in modifying the design of an airplane in order to achieve a greater cruising speed. We might begin by specifying more powerful engines for the aircraft. However, more powerful engines will weigh more than the ones originally used. This will require, among other things, that the engine mounts that attach the engines to the wings be enlarged to support the additional weight. The combined additional weight of the engine mounts and engines, though, will require that we redesign the wings to provide greater lift. The wing redesign, in turn, may force us to redesign the fuselage at the point where the wings are attached. This will add even more weight to the airplane, and the combined additional weight may force us to contemplate using even larger engines in order to achieve the desired cruising speed. Unfortunately, these engines will weigh more than the ones previously contemplated and may entail further structural changes to the aircraft.

Each step in the design process requires, in addition to new drawings, a host of detailed, time-consuming calculations. Such attributes of the product being designed as its weight, strength, and manufacturing cost must be ascertained to determine whether successive iterations of the design meet the required specifications. Thus, the savings achieved by using computers to produce and revise engineering drawings, though significant, were overshadowed by the potential savings to be gained by using computers in the design process itself.

Today, computer assisted design(CAD) is gaining widespread use. Because of the computer's great speed in performing calculations, it is possible to design and test numerous variations on a design in the computer before prototypes are built, physically tested, and accepted or discarded, as the case may be.

While CAD was being developed, similar changes were occurring in manufacturing. Initially, machine operators who produced identical or similar parts in large batches were replaced by machines that could be operated by instructions punched into paper tapes. It was just a short step from the technology used in these so-called numerically controlled machines to the technology that permitted computers to operate lathes, milling machines, drill presses, and the like in much the same way that they control robots today. Thus, computer assisted manufacturing(CAM) was born.

The next logical step was to join these two innovations. The data that are used in computer assisted design can also be used in computer assisted manufacturing, and it was not very long before the two technologies were linked (CAD/CAM).

A good deal more was happening in manufacturing while CAD/CAM was being developed. Computer programs that reduced inventory carrying costs and automatically ordered replenishment stocks from suppliers were in use by the late 1950s. These programs typically were employed by firms, such as wholesalers, who were concerned with managing stocks of purchased materials rather than inventories of work in process, such as are found in manufacturing organizations.

Manufacturers, similar to wholesalers, have inventories of raw materials and purchased subassemblies that must be managed. In addition, manufacturing organizations carry inventories of goods that are in various stages of completion in the production and assembly processes. The costs associated with maintaining inventories of partially completed and finished products comprise essentially the cost of borrowing the funds tied up in these goods. A rough idea of the potential savings that were available through better management of work-in-process inventories was evident in the observation that for every half hour that a manufactured part was actually being cut or formed in the average metal working shop, it spent about six and a half hours either sitting in an inventory or being moved from one place to another.[11]

Both inventory carrying costs and plant utilization were improved dramatically by the introduction of computer-based techniques. Descriptions of the sequences of operations required to produce each product and the amount of time needed to perform each operation were used by the computer to create detailed production schedules. Basically this was accomplished by having the computer work backward from actual orders or sales forecasts to the times at which each step in the process would be needed in order to have a completed product on the designated shipping date. Thus, it has become possible for firms using such techniques to reduce inventories by producing work in process only when it is needed. Other techniques, using similar data, were developed to improve the efficiency with which work is scheduled through the various machines used in the manufacturing process. The benefit of using these techniques is to increase the productive capacity of existing plant and equipment.

The Fully Automated Factory

There are other computer-based manufacturing techniques in addition to the ones that I have just sketched. Rather than mention them all, I will raise the crucial point that emerges from the foregoing discussion. That is, by linking the data bases used in computer applications such as those described so that all applications in a factory share a common store of information about the manufacturing process, and by placing all of these applications under the ultimate control of a master computer system that synchronizes them, it is possible to create a fully automated, computer-controlled factory—a factory capable of operating almost without workers.

[11]John I. Mattill, "The Coming of Automatic Factories," *Technology Review*. Vol. 77, No. 4, 1975, P.60.

This sounds a bit farfetched, I know. It is not farfetched, though—it is an accomplished fact. You do not even need to travel to Japan to see a computer-controlled factory; Yamazaki Machinery Works, Ltd. is building one in Florence, Kentucky, that will employ robots and only six humans. Yamazaki is planning to build an even more futuristic plant in Gifu, Japan, that will employ not only robot workers but robot tour guides as well, who will escort human visitors on plant tours.[12]

The use of robots in these plants is of major interest. It is possible to construct similar plants that do not employ robots, but such plants essentially would be giant, single-purpose machines. Similar to conventional automation, they could be justified economically only by extremely long production runs. The Volkswagen Beetle, for example, was produced with minor design changes in a production run that exceeded 15 million cars. Today it might be possible to justify a single-purpose, fully automated plant to produce a run of 15 million units. But who would gamble the investment such a plant would require on the hope that an event that has occurred only twice before in automotive history would happen again? [13] Once built, such a plant would be prohibitively expensive to modify if major changes in its output were mandated by frequent technological advances or changes in consumer preferences. What is wanted is a flexible, fully automated plant. The incorporation of robots and related equipment in such plants provides a degree of flexibility that enables plant output to be modified economically.

May the Better Robot Win

It is in Florence, Kentucky, and in the near future of American business, that most of the potential growth in the robotics industry will be found. The forecast of a $5 billion market for robots by 1992 is based, not on an assumption that robots will be used here and there in conventional factories to replace human workers at their tasks, but on the conviction that robots will be integrated into the designs of new factories that will employ relatively few people.

[12]*Business Week*, "Robots Get a Warm Welcome in Kentucky," *Business Week*. April 19, 1982, p. 48. A similar computer-controlled plant owned by Yamazaki operates today in Japan. The state-of-the-art technology employed in that facility attracts some 17,000 visitors a year. The decision to substitute robots for human tour guides, in this case, is not a frivolous one, given the labor costs associated with conducting so many plant tours.

[13]The production run for the Ford Model T car approached 15 million units.

The prospect of a large, fast-growing market has attracted numerous competitors to an industry that once was the sole province of Unimation. These relatively recent arrivals, with names as exotic as Advanced Robotics Corporation, Robotiks, Inc., and Mobot Corporation, are being joined by a second wave of competitors, whose names are household words—Bendix, General Electric, IBM, and Volkswagen.

The arrival of these industrial giants in the marketplace for robots foreshadows some basic changes in the robotics industry itself. For one thing, it suggests that the forecasts of growth for the industry are being treated seriously by large, sophisticated firms. The consequences are as follows.

The lack of support services for users of robots has been an impediment to growth in the present market. As late as 1980, Joseph Engelberger, president of Unimation, was advising readers that the potential uses for robots are limited by the necessity of maintaining in-house maintenance personnel, spare parts inventories, and the like.[14] The automobile and computer industries could never have grown as rapidly as they did had they not invested heavily in user support services. For example, IBM ran schools that trained their customers' employees to use IBM equipment and provided systems engineering help for the design and implementation of IBM products—and did not charge directly for these services. In addition, well-staffed and equipped repair and maintenance facilities accompanied every major branch sales office. Similar services still support IBM customers, although they are no longer free of charge. The point is that firms such as IBM, GE, and VW have both the experience and the resources needed to provide local support for users of their products. Moreover, the organizational structures required by both sales and service personnel exist today in these organizations. IBM is not entering a new business as much as it is merely adding a new product line to its existing business.

These firms also are among the most knowledgeable and sophisticated in their industries in terms of design and manufacturing expertise. The potential competitive edge that they enjoy in this regard may eventually drive manufacturing costs down, as well as driving their smaller competitors into segments of the market that require highly specialized, custom-designed equipment. Alterna-

---

[14]Joseph Engleberger, *Robotics in Practice*. New York: AMACOM, 1980, p. 93.

tively, the pioneering firms may be acquired by the giants, as in the case of Unimation, which was taken over by Westinghouse.

Advances will be made in all aspects of robot design, but the major improvements most likely will occur in the robot's sensory and control systems. Earlier I described the industry's present view that control and power systems are merely peripheral devices to the robot's arm. An alternative and more compelling view is that the robot is a system of machines, one of which—the arm—permits the system to manipulate physical objects. From this perspective, one is able to see the computer as the vital element of the system and the arm as one of a number of peripheral devices. One would also be led to attribute competitive advantages to firms such as IBM, that enjoy expertise in the design, manufacture, and implementation of computer-based systems.

This brings us to research and development, the topic of the following chapter. The basics of robot technology were developed by a handful of insightful, determined pioneers. Today, their line of inquiry is being pursued by hundreds of scientists and engineers in universities and industrial research laboratories. Unlike the pioneers in the field, the latest entrants have at their disposal some of the world's finest research and development facilities. They employ experienced staffs of scientists and engineers and can spend more on R & D than the pioneering firms have been able to earn in profits.

This leads to one last conclusion: that technological breakthroughs in the industry will occur at an accelerated pace. Basic problems in science and engineering cannot be solved merely by throwing money at them. However, in applied research, success often is correlated with funding. The space program has provided outstanding examples of this. Moreover, success notwithstanding, the ability just to conduct applied research often is contingent upon a start-up investment of some considerable magnitude. The GEs and IBMs of the world have made this investment in experience and research expertise that will now be applied to robotics.

5

Research in Robotics

VISION AND OTHER SENSE MODALITIES

A robot in the Charles Stark Draper Laboratory in Cambridge, Massachusetts, can assemble an alternator consisting of 17 parts in two minutes and 42 seconds.[1] The robot even tightens the product's screws, so that the assembly operation essentially is complete when the alternator leaves the work station. Robots such as this one probably will move out of the laboratory and onto the shop floor before the end of this decade. For the present, though, they must await economical solutions to a number of fascinating problems.

One problem for which solutions are beginning to appear is that of giving sight to robots. As I indicated earlier, today's working robots are limited to employment at tasks in which the exact location and orientation of each object they will encounter can be predetermined. This means, for example, that a robot cannot remove a bolt from a container of bolts and insert it into an assembly; each bolt in the container will have a unique location and orientation.

You and I would have no difficulty in doing this because we could see the bolts. Moreover, we probably would have little difficulty performing this task blindfolded, because our senses are integrated. Either a visual or a tactile experience with a familiar object will evoke recognition of that object. This integration of our senses, if integration is the right word, occurs in the brain. Touching the bolt without seeing it produces a mental image of its appearance; merely seeing the bolt evokes recollections of how it feels. People who suffer specific forms of brain damage lack the ability to tap the same pools of information with each of their senses. For example, such an individual when blindfolded may be able to describe the tactile sensations he or she has when handling a bolt but may not be able to recognize the bolt as such until the blindfold has been removed. The concept of integration of sensory inputs seems to be important in the field of robotics as well. Artificial vision systems can be augmented with other types of sensory input to compensate for the inherent shortcomings of the former.

Another aspect of human vision that guides the development of artificial vision systems is the ability to "see" both more and less than what our eyes encounter. I will describe briefly what I mean by this and then begin a more lengthy discussion of the visual processes that concludes on the same note.

---

[1]Paul Kinnucan, "How Smart Robots are Becoming Smarter," *High Technology.* Vol. 1, No. 1, (Sept.-Oct.) 1981, pp 32–40.

If you have a rudimentary knowledge of mechanical fasteners (bolts, screws, nails, etc.), you will be able to recognize almost any bolt that you see as a bolt, even as you will be able to recognize a bolt that is depicted in a crude sketch. You will see any bolt or reasonable representation thereof as a single example of a class of similar physical objects. This is what is meant by "seeing" more than what the eyes encounter.

At the same time, recognition of an object will evoke an awareness of the essential characteristics of all objects that can belong to a given class. Thus, while seeing a single bolt and recollecting the class of objects to which it belongs, the viewer will also perceive "boltness." "Boltness" is the minimum set of attributes that an object must possess in order to be thought of as a bolt, and it is something less than the set of attributes of the bolt in front of the viewer. Such attributes as the bolt's actual length, weight, and number of threads per inch are not essential to its "boltness." "Boltness" is less than the eyes encounter. The knowledge that only essential characteristics need to be recognized in order to assign objects to sets—that is, for example, to differentiate between bolts and nails—is fundamental to the development of artificial sight. It should be recalled, however, that for our purposes these essential characteristics are artifacts and not inherent properties of objects. That is, although there are inherent differences between living creatures and inanimate things, the differences between nails and bolts are of no importance except within the contexts of certain cultures.

Vision in Humans

Present theories of visual phenomena are fascinating but far removed from any degree of expertise to which I might lay claim. Therefore, what I have to say on this subject must be based exclusively on the work of others—particularly on the work of Ernest W. Kent.[2]

What we see when we look at the world around us is several steps removed from a direct experience of that world. For one thing, visual experience is stimulated by light that is reflected by the objects we see, and not by the objects themselves. What this means, among other things, is that the visual experience of an object is not a perfect, objective representation of the object itself.

We have learned to think of colors as inherent properties of

[2]Ernest W. Kent, *The Brains of Men and Machines*. New York: McGraw-Hill, 1981.

certain objects—of pumpkins as being inherently orange, for example. Yet color, far from being an inherent property, is a mental experience. Pumpkins reflect portions of the red and yellow bands of the visible spectrum of light. It is this reflected light, and not the pumpkins, that we experience as orange. Indeed, the pumpkin looks gray in the moonlight, not because there is insufficient light for its orange color to be seen, but because its color *is* gray in the moonlight and black in the absence of light and still other colors in infrared light and ultraviolet light. The eyes are not windows. They are receptors that focus reflected light onto the retina, where the light is converted to electrical impulses that, in turn, are interpreted by the brain.

Think of the retina, the lining on the inside of the back wall of the eyeball, as a screen onto which light passing through the eye's lens is focused. This screen is composed of tiny receptors that convert light to electrical signals. For our purposes, the retina can be imagined as a rectangular screen 2000 receptors high and 5000 receptors long, which amounts to 10 million receptors.[3]

The brain interprets the light patterns that form on the retina about 10 times each second. If information from each of the 10 million receptors were fed serially into the brain, as information is processed by a computer, the brain would not have enough time to process the patterns of light. You would, in this case, see a rock being thrown at you after it had hit you. Unlike present-day computers, which process data serially, the brain processes data in parallel as well as serially. Thus, although a modern computer can perform logical operations, which must be performed step by step, much more quickly than we can, it cannot process masses of data in parallel operations. For this reason, robots may never see as well as we do.

Nonetheless, 10 million pieces of data still are far too many for the brain to interpret at a rate of 10 times a second. Fortunately, nature has endowed the brain with two techniques that greatly reduce and simplify its work. First, the brain interprets the electrical signals from only those receptors that lie on the *boundaries* between areas in which the quality of light falling on the retina is different. Between one boundary and the next, the brain merely extrapolates a constant quality of light. Think of brightness as a quality of light.

Imagine, for example, that you are standing directly in

[3]The eraser on my lead pencil is about $1/4$ inch in diameter. If each receptor in the retina were the size of a pencil eraser, the screen, or retina, would be 104.16 feet long and 41.6 feet high, or 4,340.2 sq. feet. An average three-bedroom house contains about 1700 square feet of floor space, which means that the screen would be about the size of the floor space in $2^1/_2$ three-bedroom houses. Yet in actuality the retina, if laid flat, would not cover a silver dollar.

front of a huge wall, and that the half of the wall to the left of your nose is painted black, and the area to the right is painted white. A 2000 × 5000 receptor eye could "see" the wall by interpreting the signals from just 4000 receptors, rather than 10 million. The 2000 receptors to the left of the dividing line would register black. Because the wall is large, it extends beyond your field of vision. Thus, a second boundary is not detected to the left of the dividing line, and the brain merely extrapolates black across the field of vision left of center. The other 2000 receptors register white, and white is extrapolated across the field of vision to the right of center. You can become aware of this phenomenon by recalling that each eye has a blind spot where the optic nerve attaches to the retina. Yet, rather than experiencing a black hole in your field of vision, you experience vision even in the spot where you are blind. This is because the brain extrapolates and fills in the gaps.

Actually, the human eye registers two qualities of light— intensity(brightness) and color. I have not discussed color for the reason that, except for humans and a few of the higher primates, all other creatures that can see apparently lack well-developed color vision. Color vision seems to have had survival value in the process of human evolution, but it probably will not be a quality of much importance to designers of artificial vision systems for robots.

The second technique employed by the brain to reduce its data load is to attend only to *changes* in the quality of light that falls on the receptors. The survival value of being constantly alert to change(i.e., motion) seems obvious. But how are stationary boundaries observed if the brain attends only to changes in light patterns? The answer is something I observed as a young man but failed to comprehend for many years thereafter.

I once dated a girl who had pretty hazel eyes. One day, as I looked deeply into her eyes, I noticed that they "jiggled" rapidly from side to side. This was disconcerting but only momentarily so. I remembered this 20 or so years later when I learned what may, or may not, serve as an explanation.

The human eye constantly oscillates at a frequency of about 10 cycles per second. As the eye oscillates, the boundaries between areas of light of different intensity are moved back and forth across a small portion of the retina. As a boundary passes from one receptor to the one next to it, the intensity of light falling on each of the receptors changes. Consequently, the electrical signals produced by the receptors change and are interpreted by the brain.

I want to make clear at this point just what has been described and what has not. I have merely sketched out Kent's description of the process by which the eye converts patterns of light and dark to electrical signals and the techniques used by the brain to simplify the task of monitoring these signals. I have not considered the many other functions of the visual apparatus such as perception of color, depth, motion through time and space, and the like. Moreover, I have ignored other functions of the visual system such as the one that maintains a stable image while the eyes are moved. In fact, all I have described, at this point, is little more than the conversion of light to electrical signals. But this is an important subject for our purposes. Remember that the words you are now reading are encoded as electrical impulses that travel along the optic nerve between the retina and the brain—that the visual experience takes place in the brain, and not in the backs of the eyeballs.

What happens when these electrical signals reach the brain is not very well understood and usually is explained metaphorically in the following manner. Electrical impulses that travel to the brain via the optic nerve are scanned for familiar patterns. Vertical lines, for example, will generate a particular combination of electrical impulses, and this pattern is recognized by the brain as a vertical line. In a manner of speaking, your brain has a vertical line detector that will respond to this pattern of impulses by interpreting it as a vertical line. The brain's ability to convert impulses from the retina to meaningful images is, for the most part, learned.

Popular books on psychology as well as textbooks often contain reproductions of photographs or drawings that produce unusual visual effects. Among these so-called optical illusions are pictures that, at first glance, appear to be meaningless. An accompanying text usually describes the unrecognizable image in terms of the thing it represents, say a puppy, and tells the reader how to make sense of the image by pointing out the puppy's ears, tail, and so on. An image of this sort is provided in Figure 5–1.

Notice that Figure 5–1 appears to be little more than a col-

FIGURE 5–1
Visual Stimuli at the Threshold of Recognition
What is pictured? Some typical guesses are: frosty the snowman, a puppy and a leopard. It is none of these.
A description of the picture is provided in footnote 4.

lection of randomly shaped spots until one is taught to see a pattern among the spots. The pattern, of course, is a horse and rider.[4] Figure 5–1 seems to be at a threshold of perception—at the borderline, that is, of generating impulses that will trigger your "horse detector."[5] Once you see the horse, however, your "horse detector" will be activated, and you will have no further difficulty in seeing it. This is another example of what I mean by the eyes seeing more and less than they encounter. Figure 5–1 evokes both "horseness," the essence of horses, and a good deal of what you have *learned* about horses in general.

One final point must be made before moving on to a discussion of artificial vision systems. That is, that one's perception of a horse and rider brings to mind a number of mental models that give meaning to the concept of horse and rider. Seeing the horse evokes recollections of the horse-related knowledge that one has acquired, as well as recollections of knowledge that is related to horses only in some remote sense. Figure 5–2 illustrates a few of the associations that the image evokes in my own mind.

What is important to note is that a vision system can function usefully only if it evokes a model of the world that lends meaning to what is seen. In humans, the models that can be evoked seem to be interconnected, permitting fast recall of a great variety of memories. Note, for example, that the model in Figure 5–2 suggests that a horse's hoof is similar in composition to a human fingernail, as is the horn of a rhinoceros. All three emanate from hair-like cells. This type of association of memories illustrates a rich model of the world within which the concept *horse* can be understood.

The model of the world that is required by an artificial vision system will be barren in comparison. Nonetheless, that model must include some rudimentary information about the environment within which the system will operate. It may include, for example, specifications of parts to be assembled, the order in which the parts go together, the proper orientations of the parts with respect to one another in the assembled product, and so on. It is evident that artificial vision requires some form of machine intelligence. These topics will be examined in turn.

[4]The horse is facing to the left with its right foreleg raised. Most people can recognize the horse and rider after a minute or so. However, a small minority of normal people never see the horse. If you have difficulty, ask a friend to sketch in the outline of the horse for you.

[5]There is some speculation that certain drugs produce hallucinations by triggering "detectors" in the absence of the visual stimuli that normally trigger them. Thus, the hallucinator is able to "see" things that have no counterpart in objective reality. The visual hallucination can be taken as further evidence that vision is a mental experience.

FIGURE 5–2 Associations with a Horse and Rider

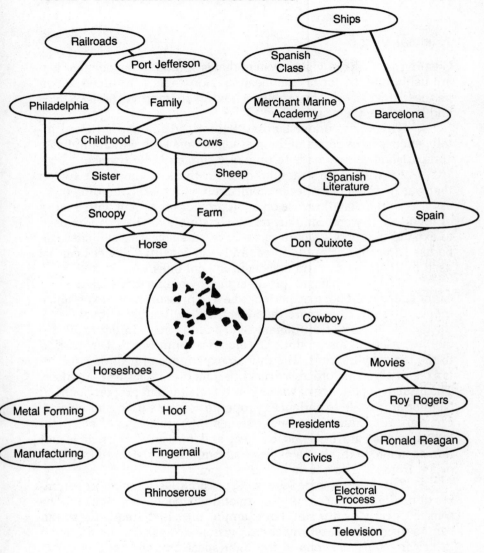

Artificial Vision

Creating an artificial vision system that is similar to the one that you are using now is out of the question. Nonetheless, artificial vision has been developed. The problems now are to improve these systems and to reduce their costs.

The key to understanding artificial vision and artificial intelligence is to understand them as metaphors with which implicate comparisons can be made to human vision and intelligence.

By way of an example, one can think of the feeler switch that was mentioned earlier in connection with plastic molding processes as a crude tactile sense organ. It provides the robot with a sense of touch, as it were. Similar, relatively simple devices can be used to expand the robot's tactile senses. Pressure-sensitive switches can be used to sense gripping force, and heat-sensitive devices can be used to distinguish warm objects from cool ones.

One may object to this metaphor by noting that such devices cannot permit a robot to experience heat or pressure as we experience them. After all, little more than electrical switching is involved in the use of this sort of equipment. The point is that, similar to vision, tactile sensations are produced by electrical impulses that are carried to the brain by neurons. The human experience of heat or pressure is a result of mental processes that interpret these electrical impulses. The human experience of warmth is not an inherent property of heat. Neither are tactile experiences properties of the surfaces of objects. They are the human nervous system's interpretations of electrical impulses that are triggered by heat and pressure. Thus, it is not unreasonable to think of a device that can distinguish hard from soft or hot from cold as a form of sense organ.

This leads to some interesting possibilities, for there are a number of media that can convey information to machines that cannot be sensed directly by humans. For example, ultra high frequency sound can be used to measure distances; vision systems can be designed to use invisible portions of the light spectrum; and magnetic and conductive properties of metals can be sensed directly by machines. The medium through which a robot senses its environment can be determined by its designer rather than by the course of evolution.

Early uses of data transmitted by visible light permitted robots to make simple distinctions between objects. However, similar to the use of limit switches as control devices, these applications were more or less task-specific. For example, light sources and photoelectric cells were used to determine the presence and orientations of parts that were manipulated by robots. If a robot was required to pick up a part that had bolt holes in it, an arrangement of lights on one side of the part and photoelectric cells on the other side could be used to determine whether or not the part was properly oriented. If it was, each photoelectric cell would detect a light that was aimed to shine through one of the bolt holes. If the part's orientation was incorrect, the lights would not be aligned with the holes and the photoelectric cells would not be activated. If the part was aligned correctly, the robot would continue in its task; if not, the robot would discard the part and grasp another before proceeding in its work.

A more sophisticated use of light was developed to permit robots to make distinctions between different objects. In this case, objects were moved by a conveyor belt past a focused light source. A photoelectric cell opposite the light source would detect objects as they interrupted the light by moving between the light and the photoelectric cell. The ability to make distinctions between different types of objects was developed in the following manner. Since the conveyor belt moved at a constant speed, objects of equal size would block out the light for equal time intervals; an object that was two inches long would interrupt the light for two seconds if the belt on which it moved traveled at a rate of one inch per second. Thus, a robot could be programmed to recognize objects according to the amount of time they took to pass in front of the robot's "eye."

The device that I have described was much more complex than it sounds. For one thing, it incorporated the rudiments of an "object" detector. A two-second interruption of the light source, for example, produced an electrical signal that was interpreted as a particular kind of object. Moreover, a world model of sorts was incorporated in the device. In order for the system to work, the robot had to know, for example, that a two-second interruption was a particular kind of bolt, and that the bolt had to be grasped in a certain way and inserted in a particular bolt hole. In a sense, the robot in this illustration knew how an object that it identified fit into its scheme of things.

There are other complicating factors to be considered. For the system under consideration to function, the objects to be differentiated from one another must have different profiles. This is not always the case. Moreover, the objects must either be properly aligned with respect to the conveyor belt or the vision system must be programmed to recognize a part when it is randomly oriented with respect to the belt. The problem is that a rectangular solid that is, say, 3.0 inches long and 2.0 inches wide will have a profile that varies from 2.0 inches in length (when its long side is perpendicular to the viewer) to 3.6 inches (when the object is cater-cornered with respect to the viewer.) How, then, can a vision system differentiate between one object that is 2.0 inches wide and 3.5 inches long and another that is 2.5 inches wide and 2.8 inches long? All possible profiles of the latter object can be formed by the former one. Even more difficult is the problem of recognizing objects that touch one another.

Such problems can be solved, but the solutions tend to make the resulting system more and more task-specific. What is wanted is not a single-purpose vision system but a flexible one. A different approach to the problem can provide the latter.

A number of manufacturers presently market devices that are known as *silhouette vision systems*. Such devices have operating characteristics similar to human vision. In principle, they work as follows. A television camera records the objects that appear before it. The camera is set so that it records only in black and white, and not in gray. Thus, if colored objects are placed on a black background before the camera, the objects will appear as white silhouettes against an otherwise dark field of vision. Above a certain, predetermined value, all shades of color will be converted to white; all shades below that value will be recorded as black.

The data that are recorded by the TV camera are transmitted directly to a computer, much as electrical stimuli that are produced on the retina travel to the brain. The computer becomes the "mind's eye." In much of the discussion that follows, I will attempt to visualize the operations that the computer performs on these data by imagining that the computer looks at a TV screen on which the data are displayed. Although a television set can be attached to the camera to provide human operators with a view of what the system "sees," the computer actually deals directly with the electrical signals that are sent to it. As far as the computer is concerned, there is no TV set.

The data received by the computer are scanned for edges—an edge is assumed to occur whenever a signal representing black is followed by the signal for white and vice versa. Each of the scan lines from the TV camera is analyzed in this way, and as successive scan lines are analyzed for edges, the perimeter of each object in the camera's field of vision begins to take shape in the computer's memory. A complete perimeter is recognized when points representing edges on former scan lines are not followed by contiguous points representing edges on subsequent lines.

When a complete perimeter of an object has been received by the computer, it is stored temporarily as an electronic model of the object's silhouette—that is, as a string of data that are encoded as high and low voltages, for example. The computer will have in its long-term memory similar strings of data, or models, that represent the different objects that it has been programmed to recognize. Thus, the perimeter of an object before the camera is compared with the perimeters of the objects for which the computer is looking, and recognition of an object occurs when a perimeter that has been recorded by the camera matches one in the computer's memory. A perfect match is not required—an approximate match will permit the computer to recognize an object. An analogous, close-matching process in the human vision system permits us to recognize familiar objects when they are seen indistinctly.

A review of the artificial vision system as it has been described so far will reveal a number of parallels to human vision, as well as the system's inherent flexibility. Similar to human vision systems the robot's system looks only for edges. It creates patterns of electrical impulses that trigger "pattern detectors." The pattern of impulses associated with a particular object, or a reasonable approximation of that pattern, is taken as the essence of that object. The visual essence, moreover, is learned through prior experience which, in the case of a robot, takes the form of programming.

The system's flexibility is apparent in its generality. The ability to recognize one object or another is not a function of the system's hardware, but of its programming. As you may have concluded, the system will recognize objects regardless of their orientation and location in the plane of the camera's field of vision. Obviously, an object's perimeter is independent of its location on a plane.

Nonetheless, the robot ultimately must determine both the location and orientation of an object that it has recognized if it is to be able to manipulate that object. How this is accomplished is fairly simple. Imagine that the computer recreates in its memory an analogy of what lies in the camera's field of vision. Along the left edge of the field the computer superimposes the vertical axis of a graph. The bottom edge becomes the graph's horizontal axis. The lower left-hand corner of the field of vision, then, corresponds to the origin of the graph.

Next, the computer calculates an ellipse that has the same area as the object in its field of view and substitutes this ellipse on the graph for the object. The ellipse is located so that its geometric center coincides with the geometric center of gravity of the object's silhouette. Having done this, the computer will be in a position to approximate the location and orientation of the object in question; its location is calculated as the geometric center of gravity of the silhouette, and its orientation as the angle formed by the major axis of the elipse and the horizontal axis of the graph. A pictorial representation is shown in Figure 5–3. Recall, however, that the computer does not draw pictures such as Figure 5–3 or look at a TV screen—it merely calculates the mathematical equivalent of what I have illustrated.

The system that has been described is capable of recognizing numerous objects that appear simultaneously in its field of vision. Moreover, it can be programmed to recognize objects that move on a conveyer belt. It can be instructed to ignore, temporarily, objects that are not entirely within its field of vision. This can be accomplished by programming the computer to ignore what it sees whenever its field of vision does not include an unbroken black

FIGURE 5–3 An approach to Estimating the Location of an Object Using Artificial Vision (Adapted from: Paul Kinnucan, "How Smart Robots are Becoming Smarter," *High Technology*. Sept./Oct., Vol. 1, No. 1, 1981, p. 35.)

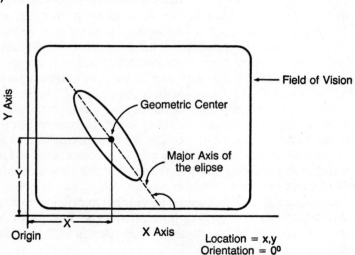

border: an object that is only partially within the field will create a white gap in the border. Other problems, such as the identification of objects that are touching or overlapping, await elegant solutions.

One final point will lead us to a discussion of artificial intelligence. As I noted earlier, vision implies a world model. Thus, the robot must incorporate in its program a model that defines the objects that it recognizes in terms of what the robot is to do with each object and when it is to do these things. For example, the robot must recall that a particular bolt that it has recognized is to be grasped at a certain point with respect to the location of the elipse's geometric center and inserted in a particular part of an assembly. It must also recall the order in which this step is to occur, since the arrival of parts in the robot's field of vision may not correspond with the order in which they are to be used by the robot.

ARTIFICIAL INTELLIGENCE

What is intelligence? For all of the effort that psychologists have devoted to the measurement of intelligence, they have yet to produce a general definition of the term. In fact, psychological testing of intelligence can result in a tautology—that is, no operational definition of human intelligence exists apart from the performance on these tests of the human subjects who have taken them. Any common-sense notion of intelligence must incorporate more in the way of a definition of mental prowess than an aptitude for taking tests. After all, there are apparently bright people who routinely perform poorly at such tests, and there are others who score highly but behave in everyday affairs as if they were not particularly smart.

There is no well-defined concept of human intelligence. Thus, I must make my position on artificial intelligence clear at the start. If we cannot define intelligence in humans, we certainly cannot decide that computers can become as intelligent as people are. Nonetheless, it appears to be an article of faith in some quarters of the field that machine intelligence will become coextensive with human intelligence. There are many reasons, which I cannot elaborate here, that lead me to reject this notion.

There are machines that can perform tasks that, hitherto, only humans could perform and that only smart humans could perform well. There are, for example, chess-playing computers that can beat all but a handful of expert human chess players. My position is that such machines display machine intelligence because they perform tasks that require intelligence of the humans who perform those tasks. That, however, is all I take the term *artificial intelligence* to mean. The ability to play chess well or to score highly on an IQ test merely is a demonstration of intelligence—yet, neither is coextensive with intelligence or a perfect correlate of mental ability. Nonetheless, machines can behave intelligently.

My position, though subject to debate, is based on the existence of machines that can play chess and do other interesting things. Thus, it is not a difficult position to take. However, the issues regarding artificial intelligence actually arose before the development of such machines.

Work on artificial intelligence predates even the early vacuum tube machines. One of the pioneers in the field of artificial intelligence, Alan Turing, proposed a simple test that he imagined could be used to determine whether a machine is intelligent. The Turing Test goes something like this. An interrogator, who cannot see the one he interrogates, is asked to guess whether the one being interrogated is a human being or a machine. Machines that can be interrogated by humans and that are mistaken for humans by their interrogators are, by Turing's definition, intelligent. That is an elegant definition. You will note, though, that it still does not tell us what intelligence is. Far worse, the definition may permit human gullibility to decide the issue for us. An example of what I mean is as follows.

Some of the research on artificial intelligence resulted in a family of programs that simulate human verbal responses. These are interactive computer programs that produce dialogs with human respondents. One such program, *Doctor*, simulates a session with a nondirective psychologist, in which the computer is the doctor and the human is the patient. A sample of typical output from a session with *Doctor* is provided below:

HUMAN: I don't get along with my brother Bill.
COMPUTER: *Tell me more about your family.*
HUMAN: There's five of us altogether. Bill is older.
COMPUTER: *Please go on.*
HUMAN: Even as a kid Bill always ignored my rights.
COMPUTER: *Your rights.*
HUMAN: Yes. He never gave me a break.
COMPUTER: *You seem quite positive.*[6]

[6]John Krutch, E., *Experiments in Artificial Intelligence for Small Computers.* Indianapolis. IN.: Howard W. Sams & Co. p. 86.

Before we conclude that *Doctor* is an intelligent program—one that could pass Turing's Test with flying colors—it must be pointed out that there is less to *Doctor* than meets the eye. *Doctor* knows nothing of personality theory, sibling rivalries, or psychological decompensation. In fact, it might be fair to say that the program is a trick. The trick is that *Doctor* contains a glossary of key words. Each line entered by the human subject is scanned for these words. In this case, *brother* is a key word. The word *brother*, as well as *sister*, *mother*, and *father*, will automatically trigger the stock response, "Tell me more about your family." Now, if you substituted the word *jackass* for *brother* in *Doctor's* glossary, and then began the interview by saying, "I don't get along with jackasses," *Doctor* would reply, "Tell me more about your family."

Variations on *Doctor* include *Parry*. *Parry* is a patient with the symptoms of paranoid schizophrenia whose dialogues are somewhat more convincing than are *Doctor's*. Occasionally, *Doctor's* replies are a bit off-the-wall, as the saying goes. *Parry's* off-the-wall conversations, if anything, appear to be further evidence of schizophrenia. If you're wondering whether *Parry* and *Doctor* can "talk" with one another, they can.

In any event, anyone can be fooled, and that is what I see to be the fatal flaw in the Turing Test. At the time that *Doctor* made its debut, certain mental health practitioners proclaimed that, once *Doctor* was refined, unlimited psychological counseling could be made available to anyone who had access to a computer terminal.

As I have indicated, *Doctor* only mimics the responses that one would expect to hear in a Rogerian therapy session. Yet, some mental health professionals, as well as laypersons, attributed to the program a potential to acquire the vast store of knowledge, the understanding of the human condition, and the skills with which these are used in therapeutic sessions that exemplify competancy in this profession. This sort of blind, uncritical faith in technology, which has parallels in almost every technological endeavor, became a matter of considerable importance for *Doctor's* creator, Joseph Weizenbaum. Weizenbaum reasoned that a therapist's skill in helping patients to resolve their emotional problems ultimately rests in the therapist's experience with and eventual understanding of such problems. He was disturbed, then, to find that some professionals, in their enthusiasm for *Doctor*, were suggesting that such understanding could be "—supplanted by pure technique." Perhaps disturbed is too gentle a word to describe Weizenbaum's reaction, for as a result of this experience, Weizenbaum took leave of his work in computer science to spend two years developing an essay that explores: (1) the differences between humans and machines and (2) the identification of tasks that computers ought not be assigned to perform, whether or not the development of artificial intelligence ultimately makes such assignments possible.[7]

Problem Solving

A potentially more useful way to view artificial intelligence is through the application of machines to problems that formerly only humans could solve. Much of the early research in this field was devoted to game-playing machines. Researchers began by programming machines to play tic-tac-toe and checkers, and then went on to the more challenging game of chess. Games such as these were obvious choices for the pioneers in this field; virtually everything that a computer needs to know about chess is contained in the rules of the game.

As anyone who has learned the game can tell you, the rules are fairly simple: my children were playing chess when they were 9 and 10 years old. Mastering the game is quite another matter. Good chess playing requires, among other things, the ability to visualize the numerous consequences of each possible move for six or seven moves in advance of the move in question. The thought processes involved result in a branching network of consequences, with each potential consequence evaluated in terms of its effect on the player's resources and strategy.

Computers are lightning fast when it comes to performing sequential operations such as those required in chess playing. Yet much more is required of a chess player than this. The number of possibilities in a game of chess—that is, the number of possible consequences that can arise in a game—are estimated to be in the neighborhood of 10^{120}. The equivalent form of that number is 10 followed by 120 zeros, which is a number somewhat larger than the estimated number of molecules in the world.[8]

Human chess players obviously cannot consider so many alternatives; neither can computers. What humans apparently do is to investigate a limited number of possibilities—just those that seem

[7]Joseph Weizenbaum, *Computer Power and Human Reason.* San Francisco: W.H. Freeman and Co., 1976.

[8]Herbert Simon, "Is Thinking Uniquely Human?" *University of Chicago Magazine.* Fall 1981, p.14.

to bear the most fruitful outcomes. Chess-playing computers also do this by limiting the number of alternatives to be explored in depth by applying what are known as *heuristic rules*. A heuristic rule is a rule of thumb such as: "It is better to control the center of the board than it is to control its periphery."

Although humans and computers both employ heuristic rules to aid decision making, the ways in which they are used by men and machines seem to differ dramatically. An accomplished chess player will reduce the number of alternatives to be explored to about 100, which are investigated in varying degrees of depth within the allowed three-minute period between plays. In the same time, a chess-playing computer may investigate upward of 25 million possible moves.

Twenty years ago, it was common to hear computers described as glorified adding machines. It was claimed that they could do no more than repeat a series of instructions that were steps in solving a problem that previously had been solved by the programmer. In part, this was an apt description of the ways in which computers commonly were used at that time. It also was a reflection of the images that manufacturers presented of their machines to counter the public's mounting apprehension about machines that appeared to think. In fact, computers such as those that play chess do much more than follow rote instructions. The best chess-playing machines have attained expert ratings and often are better at the game than are the people who program them.[9] They do not proceed from start to some predetermined end but achieve ends that cannot be predicted because those ends are determined in large part by the opponent's strategy and skill.

Coping with the Real World

It is ancient history now, but at one time scientists were engaged in building maze-running machines. No bigger than rats, these machines typically would bumble their way through a maze by trial and error, but they would "remember" the sequence of turns that would permit them to negotiate the maze successfully. On subsequent trials, they would know how to run the maze without getting lost in it. In a manner of speaking, they were able to learn the maze.

[9]David L. Waltz, "Artificial Intelligence," *Scientific American.* (Oct. 1982), Vol. 247, No. 4, pp. 118–133.

An interesting footnote to this period in history concerns a little robot that could wander about a room and extricate itself from corners and cul-de-sacs. In addition, this robot monitored the amount of energy in its rechargable batteries and plugged itself into a wall socket whenever its batteries ran low. In a metaphoric sense, it got hungry and looked for a place to eat.

As the story goes, two such robots were placed in a room that had only one wall socket. They bumbled around for a while and eventually began to search for the socket as their batteries became depleted. One should not make too much of what happened next, but it is difficult not to do so. The robots found the wall socket at the same time and began, simultaneously, to attempt to plug themselves into the single receptacle. As they bumped into one another, backed off, and collided again, they mimicked a struggle for survival. Eventually, one of the robots ran out of juice and died, as it were, while the other plugged itself in and lived to bumble another day.[10]

An Infant's World

Despite their lack of sophistication, machines of this sort displayed two attributes that one would expect of an intelligent being. They modified their actions in light of their experiences, and they were able to operate in the real world where the rules are not all spelled out in advance.

These two attributes, along with the kind of decision-making ability that is apparent in game-playing computers and the ability to process natural language that is employed by *Doctor*, were combined in a program called SHRDLU.[11] SHRDLU sits before a pile of blocks of different shapes, sizes, and colors, which it is able to manipulate with its robot arm. A typical session with SHRDLU might go something like this:

HUMAN: Place the blue pyramid on top of the red block.
COMPUTER: *OK*
HUMAN: Did you touch any blocks beside the blue one?
COMPUTER: *Yes. A green block was on top of the red block. It was removed so that the blue pyramid could stack on the red block.*

[10]Instructions for building a similar robot are contained in the following source: David L. Heiserman, *Build Your Own Working Robot*. Blue Ridge Summit, Pa.: Tab Books, 1976.

[11]S,H,R,D,L, and U are the seventh through the twelfth most frequently used letters in the English language.

The following points about SHRDLU should be noted carefully: (1) The program communicates with its human operator in natural language, (2) it is able to abstract the meanings of terms that are not defined explicitly (for example, "blue one" must be identified by the program as a blue block, and, more precisely, as the blue pyramid that the computer touched), (3) the logic used by SHRDLU is "back-chained"—that is, it starts with an end goal such as blue pyramid atop red block and works backward through the steps (i.e., remove green block from red block) that must be achieved in order to get from the initial conditions to the end goal, (4) the program is able to recall and reconstruct the logic that leads to its actions, and (5) it deals with objects that have physical properties—that is, for example, with pyramids that cannot be stood on their apexes.

Unlike *Doctor*, which is ignorant (one is tempted to say, *who* is ignorant) of the meanings of the words it employs, SHRDLU has a basic understanding of the world of blocks in which it operates. Subsequent experiments with SHRDLU have refined it in subtle ways. For example, the program has gained the ability to learn—it can generalize from problems such as the one in the previous illustration, commit the generalized instructions used to remove one block from another to memory, and recall them each time the need arises. Another refinement has been the addition of the capacity for processing parallel information. As SHRDLU delineates the alternative steps that might be taken to execute an order, a subprogram examines each alternative step and criticizes it. When such criticism uncovers problems with a step, a second subprogram comes into play to rectify the problems thus encountered. The processes of alternative generation, analysis, and revision occur simultaneously, mimicking the parallel processes that occur in the human mind.

One final point about SHRDLU. There is no robot arm or set of blocks.[12] There could be, but it would be gratuitous to provide them. The blocks, their sizes, shapes, colors, and locations, are represented merely as strings of data in the computer's memory, as are the robot arm, its position, motion, and the like. The world of blocks exists in the mind's eye of a robot.

I find these images to be compelling. The blocks suggest an infant's playthings—the fact that they exist, not as physical objects but as abstract symbols, suggests an intelligence that is capable of seeing laws and generalizations in the apparent chaos of experience.

[12]David L. Waltz, *op cit.* p. 120.

It is moving and, indeed, awesome to speculate on the abilities and eventual uses of SHRDLU's descendants—of the third and fourth generations, as they are called, of artificial intelligence technology.

The World of Work

It may be some time, still, before SHRDLU takes its place on the factory floor to relieve us of our toil. One need not wait at all, though, to find artificial intelligence in other forms joined in human endeavors. The places in which you will find artificial intelligence applied to human problem solving are hospitals, laboratories, and research facilities. The backward chaining process used by SHRDLU has been incorporated in programs that aid physicians in the diagnosis of disease. Other programs, equipped with elementary mathematical concepts, generate and test new mathematical theorems. Scientists are assisted in their laboratories by programs that perform essential tasks that formerly were assigned to laboratory assistants.

Machine intelligence has passed the point of humble beginnings and looks to be a vital element of future generations of technology. All new technologies seem impractical at one point or another in their early stages of development. Television was deemed to be an interesting, but impractical, development just over 50 years ago— it took a staff of technicians just to keep a TV receiver functioning. The first diesel engines used powdered corn cobs rather than oil for fuel. So too were the little, bumbling mechanical rats impractical laboratory gadgets, or so it seemed. In contrast, intelligent computer programs, robots, and the like are gainfully employed today—they have emerged from the state of being mere scientific curiosities. Yet, it is simple to overlook this fact. When is the last time you saw a robot or played chess with a computer?

6
Robots and Organizations

Technology and Culture

People of exceptionally high intelligence characteristically become curious about many things to which they apply their extraordinary talents. The true mark of such intelligence is to have made significant contributions not to one field of human endeavor, but to many. Thus, Erwin Schrödinger, the physicist who discovered wave mechanics, also involved himself with problems in such diverse fields as philosophy, cellular biology, and the human mind.

In a series of lectures on mind and matter, given some 25 years ago, Schrödinger noted that the evolution of a species requires something more than the genetic inheritance of advantageous physical traits.[1] The evolutionary value of a physical trait, he reasoned, often is inextricably linked to parallel changes in the behavior of the organism. Neither sharp claws nor swift limbs enhance an organism's probability of survival by themselves: survival is made more likely only if the genetic change that produced them is accompanied by an instinctive urge to use these new traits in survival-enhancing ways. "To distinguish between the possession of an organ and the urge to use it and to increase its skill by practice, . . ." Schrödinger argued, ". . . would be an artificial distinction made possible by an abstract language but having no counterpart in nature."[2]

I have raised Schrödinger's thesis in order to introduce a corresponding view of the effect of technology on culture: namely, that the evolution of technology inevitably must be evaluated in the context of the cultural changes that it affects. Just as the prehistoric reptiles that evolved into flying lizards became quite different creatures from ordinary lizards as a result of their predisposition to fly, so a culture can be transformed into a different culture by its urge to use its inventions. Unfortunately, not all such changes seem to be for the better.

Weizenbaum illustrates this point in the following way.[3] According to the popular view, the invention of computers occurred just in the nick of time. One cannot imagine, for example, how the

[1]Erwin Schrödinger, *What is Life? & Mind and Matter*. London: Cambridge University Press, 1967.

[2]Erwin Schrödinger, *op. cit.* 1967, p.121.

[3]Joseph Weizenbaum, *op. cit.* 1976.

welfare distribution system would have been able to continue functioning through the 1960s and 1970s had not its work been transferred to computer systems. Another view, Weizenbaum's, is that
the introduction of computer technology may have prevented much-
needed reforms in the welfare apparatus. In the absence of computers, the mounting size and complexity of the problems encountered in the welfare distribution system might have forced the adoption
of innovations, such as a negative income tax, that would have elim-
inated the need for an elaborate bureaucratic apparatus. Instead, the
introduction of computers permitted the existing system to survive
and, eventually, to grow. Growth, accompanied by a corresponding
increase in the political power of the welfare system, rendered subsequent innovations, such as the negative income tax, less likely to
succeed as viable alternatives to the welfare bureaucracy. We all may
be the worse off for that.

Science and Culture

Weizenbaum's point is speculative and anecdotal; it was not his
intention to define rigorously the range of effects that technologies
can have on organizations. Others have done this, and what follows
is a summary of their conclusions. By way of a preface to this summary, however, I want to comment briefly on the validity of knowledge that is gained from scientific endeavors.

The validity of all scientific knowledge is problematical. It
is important to know this, but it is also sometimes essential to ignore
it, as I will explain.

It often seems that familiar "scientific facts" are absolutely
and eternally true, but the history of science suggests that this has
never been the case and may never be so. For example, it seemed
certain that time, however measured, was constant everywhere—
that a second would take as long to pass here in my office as it would
everywhere else in the universe. We know now that this probably
is untrue. Einstein's strange-seeming, relativistic theories, together
with corroborating experimental evidence, lead us to conclude that
time is relative—that a second in my office will consume a different
fraction of my lifetime on earth than that second would of your life
if you were on a spaceship, traveling close to the speed of light. Your
clock would run more slowly than mine would.

It is important to remember that the experimental evidence
marshalled in support of any "scientific fact" can never be sufficient

to guarantee that that "fact" is unconditionally true and that it will never be replaced at some future time by a "truer fact." It is equally important to know that the experimental evidence gathered in support of any scientific fact is never unequivocal in its totality. In the course of scientific endeavor, evidence is marshalled both for and against accepted scientific facts. The evidence against is ignored, so to speak, until a newer, competing theory emerges to revise or redefine the facts.

One must be aware of this in order to understand how the scientific enterprise functions. However, one also must be willing to ignore it in the course of everyday activities—regardless of whether these activities are pursued in the name of science, education, or commerce. Although the validity of scientific knowledge is problematical, one can use only the knowledge that is at hand. Moreover, to use such knowledge effectively, one must be prepared at least to act as if its validity was beyond question and, in many cases, to believe this to be the case as well. To put it more concretely, a scientist cannot devote a lifetime to a field of study while doubting the scientific theories upon which that field is based—neither can a business executive be kept from acting by the knowledge that such actions take place on shaky ground.

Fortunately, progress results in spite of everything because it is possible to base fruitful actions on invalid facts. Useful techniques for navigation and surveying, for example, were derived from the invalid pre-Copernican conception of an earth-centered universe. Theories and scientific facts, even when invalid, encourage us to pursue useful thoughts and actions that we could not have contemplated in their absence.

I have belabored these points because of my reservations about the validity of some of the ideas that follow. Curiously, these reservations are accompanied by a conviction that, nonetheless, the ideas provide genuinely useful ways to think about and manage organizations.

Technology and Organizations

The word *technology* brings directly to mind thoughts of gadgets and machines. A more abstract, but potentially more useful, thought is one of technology as the means that are employed in solving problems. This broader definition permits us to view a computer terminal and a computer program as different examples of technology. The former is a machine that represents a solution to the problem of finding a medium through which one can communicate with a computer. The latter is a list of abstract symbols that constitutes a solution to the problem of transforming a universal (general-purpose) machine into one that can perform a specific task. Moreover, it is a set of instructions that, in theory, a human could follow in order to perform the same task—although, in actuality, the relatively slow rate at which humans process information serially may render it impossible for any of us to complete some of these tasks in the space of a lifetime.

The definition that I have proposed allows us to differentiate among technologies according to the quality of the solutions to problems that they provide, rather than according to their design characteristics. This means, for example, that an automobile's ignition system and an electroshock machine that is used to treat cases of chronic depression will be seen as different technologies rather than as similar ones that produce timed, regulated electrical impulses. The point is not that these technologies are used to solve quite dissimilar problems, but that the quality of the solution provided in each case is quite different.

The technology exemplified by the ignition system—the solution to the problem of igniting gasoline—is *well structured*. By well structured, I mean that the problem, and consequently the technology, is well enough understood that it lends itself to logical, structured analysis. One can explain how and why the technology works, using accepted "facts" from basic physics and chemistry. Moreover, one can use these same facts to modify the technology so that it lends itself to numerous, different applications, as would be the case, for example, if an automotive ignition system were to be redesigned for use in a tractor.

The technology used in electroshock therapy, in contrast, is not well understood—it is *ill structured*. It is not clear why some individuals suffer from chronic depression, and it is even less clear why electroshock therapy works in some cases to improve their condition. Although there is a logical, structured answer to the question of how the machine produces electrical impulses, there is no such answer to the question of how these impulses intervene in the process of mental illness.

Basically, problems that can be solved using well-structured technologies can be "solved" on paper before they are solved in the field. The problem of writing a program that will instruct a computer to produce an organization's operating budget can be thought out in advance and, essentially, solved long before the programmer goes near a computer. Such problems allow a planner to anticipate contingencies and to plan solutions for them.

The problems to which ill-structured technologies are applied cannot be solved in the abstract—they can be solved only as they are encountered. Solutions, when they can be found, invariably seem to rest on such intangible human capacities as judgment, insight, and creativity. One cannot plan in advance a set of procedures that will result in the discovery of new scientific theories, for example. Progress in science seems to result, as much as anything, from trial and error coupled with keen insight.

What follows is an illustrative list of tasks that are relatively well- and ill-structured. They are intended to represent extremes on a continuum rather than discrete categories of technologies.

A second, simpler way to view technology is according to the extent to which the problems to be solved vary. At one extreme, in the case of machine tool operation, a single part may be made repetitively. At the other extreme, the same type of metal-working machine may be employed in a shop that produces one-of-a-kind items for a variety of customers.

Taken together, these two ways of describing technologies provide a table with which differing technologies can be classified. The table, which is illustrated in Figure 6–1, can be interpreted as follows:

The corners of the table describe what can be imagined as extreme types of technology. The lower left-hand corner, which represents the application of well-structured solutions to problems that do not vary, describes continuous-process technologies. Oil refineries, milk bottling plants, and steam generating facilities employ routine en-

TABLE 6–1. Well-Structured and Ill-Structured Tasks

WELL-STRUCTURED	ILL-STRUCTURED
machine tooling	sculpting
machine scheduling	personal selling
writing procedures manuals	writing advertising copy
engineering	basic research
payroll accounting	employee selection
writing programmed instructions	writing poetry
janitorial work	employee counseling

FIGURE 6–1.

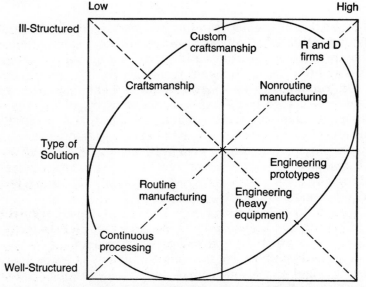

Adapted from: Charles Perrow, "A Framework for the Comparative Analysis of Organizations," *American Sociological Review*. Vol. 32, 1967, pp. 195–208.

gineering solutions to what amounts, in each case, to a single problem. The only variations in the problem of generating steam, for example, may be changes in the rate at which steam is used, changes in the quality of fuel oil used in the boilers, and the like.

The upper right-hand corner of Figure 6–1 represents the opposite extreme, where variability is high and solutions to problems are ill-structured. Research and development firms may be involved in attempts to produce materials that are stronger, lighter, and less expensive than aluminum, to develop economically viable processes that use sunlight to convert water to hydrogen and oxygen, and to invent inexpensive robot vision systems. The solutions to such problems are ill-structured. One will not find them in textbooks or scientific journals, nor will one find descriptions of the processes that must be employed to invent them. The problems encountered by such organizations vary over time as well; as one problem is solved, another must take its place.

What becomes apparent at this point is that firms as dissimilar as bottling plants and research and development laboratories will be organized and managed quite differently. The introduction of robotics to manufacturing operations probably will occur in organizations that employ the technologies described by the diagonal line that runs from the lower left-hand corner to the upper right-hand corner of Figure 6–1. For the sake of completeness, I will go on to describe briefly the remaining two corners of Figure 6–1. The remainder of this chapter, though, will provide a discussion of the technologies that lie on this diagonal line, the types of management and structure that such technologies seem to require, and the changes that robots are likely to make in these organizations.

The upper left-hand corner describes the essence of craft technologies. Here, intuition rather than science, and judgment rather than codified knowledge, inform the way in which work is performed. Skilled glassblowers, violin makers, and lithographers apply techniques that cannot be codified in formulas, rules, or procedures and, thus, cannot be transferred to machines. This is not to say that violins and glass bottles cannot be machine made, but to acknowledge that machines cannot make violins that come close to matching those of a master violin maker.

The lower right-hand quadrant describes technologies that apply well-structured solutions to a variety of problems. An engineering firm employing this sort of technology may design and build centrifugal pumps of different sizes and designs to handle different types of fluids. Regardless of the variety of design problems faced by the pump manufacturer or the degree of complexity of each design task, each design and manufacturing task has a logical solution in the sense that the calculations and procedures needed to complete each task are known by, or available to, the firm.

MANAGING ROBOTS AND PEOPLE

Bureaucratic and Democratic Organizations

As I suggested, organizations that employ technologies similar to those described by the diagonal line in Figure 6–1 will function differently, depending on where the technologies used are located on the diagonal. Organizations using technologies that lie at the lower left-hand end of the diagonal tend to be bureaucratic. Managers of such organizations are in a position to know exactly how the organization's tasks are to be performed. Consequently, they are able to create a structure that controls the way in which their subordinates perform these tasks. Work is planned step by step, and procedures are outlined for each step in the processes. Subordinates typically are not free to exercise their discretion at work but must follow the plans and procedures laid down by their supervisors.

Such organizations and the people who work in them are evaluated in terms of the efficiency with which they operate. The output of these firms often is indistinguishable from the output of competing firms. Market prices for the output tend to be extremely competitive since similar firms typically use similar technologies and raw materials. The crucial problem confronting these organizations is not *how* to produce what they make, but how to produce it competitively—how to reduce costs or to improve quality without increasing costs to uncompetitive levels.

Organizations at the other end of the spectrum, such as R & D firms, have quite different concerns. For their management, the vital issue is not efficiency, but effectiveness. The success of such firms depends primarily on whether they can achieve their objectives, and only secondarily on the efficiency with which these objectives are met. As an extreme example, one should not be particularly concerned about the cost or efficiency with which a cure for cancer is found; the vital issue is to find a cure—almost any cost will be acceptable to society.

I do not mean that such organizations as these waste resources or that they possess resources in unlimited amounts. Indeed, a firm may choose one research strategy or project over another because of cost considerations. Similarly, an organization may turn down opportunities for research projects that would exceed that organization's resources. Nonetheless, because each task facing these organizations tends to be unique, there are few, if any, opportunities to benefit from learning curves or economies of scale.

Because solutions to ill-structured problems cannot be determined on the basis of rigorous, logical analyses, managers cannot preplan the work of their subordinates to the extent that they can in bureaucracies. Many of the crucial decisions that effect work in these organizations must be delegated to the people who perform the work in question. The ability to make effective decisions in such cases often rests on the experience, judgment, creativity, and intuition of the decision maker. For a given task, these qualities usually reside in the individuals who perform that task and not necessarily in levels of management that are removed in the organization's hierarchy from that task. Moreover, the ability to make decisions with regard to a particular task often is informed by feedback—that is, by information that is gained from the task as it is being performed. Again, such feedback is available to the individual performing the task, but not necessarily to managers, who have little direct experience with the task.

The result is that organizations that use ill-structured technologies are neither managed nor organized as bureaucracies. Fewer rules and procedures exist in these organizations than in bureaucracies. The authority to make decisions is delegated further down the organizational hierarchy. Coordination of efforts among individual workers and among work groups and departments is achieved by direct, often informal, communications. In bureaucracies, such coordination frequently is effected solely through predefined channels. In essence, these organizations, relative to bureaucracies, are less structured and more flexible and democratic.

Bureaucracy and Robotics

The firms that are most likely to benefit from robotics presently employ technologies that are described by the mid-point of the diagonal line in Figure 6–1; these technologies are not entirely well-structured, and are subject to considerable variability. Production runs in such organizations are changed periodically, and each change is accompanied by a host of ill-structured problems.

Until now, the distinction that I have made between ill- and well-structured technologies has been based solely on the quality of the scientific disciplines from which the respective technologies are derived. For example, the explanations from physics and chemistry of combustion in an automotive engine are more complete and, thus, qualitatively different from the explanation of electroshock therapy that psychiatry provides. An additional distinction must be drawn. In practice, ill-structured technologies are often employed by organizations even though corresponding well-structured technologies are available to them. There are, for example, well-structured solutions for production problems such as machine scheduling. Yet, many firms do not use these technologies to their fullest extent—in some cases because they lack sufficiently trained personnel. Other firms use these technologies piecemeal in that different parts of the production management problem are handled by different departments of the organization. Consequently, solutions that are developed by different units of the organization often are not entirely compatible with one another.

Compatible, well-structured solutions to production planning problems require the use of common data bases and integrated problem-solving techniques. For example, data from sales forecasts must inform planning efforts for material requirements, which in turn must feed information into the management system for raw materials inventory. The activities of these three planning systems, in turn, will affect and will be affected by such activities as planning the work force, product design, and the like. When all such planning activities are pursued independently, the problem of coordinating them in the factory becomes ill-structured, even though a well-structured technology may have been applied in each of the individual planning activities.

The piecemeal introduction of robots to replace a welder here and a painter there will not solve the problem that I have described. The problem is addressed in the design of plants that make extensive use of robots and related equipment. The heart of such plants is a computer-based control system that integrates major production planning and control activities. It may be that only such computer-based systems are capable of making full use of well-structured technologies in manufacturing and assembly operations.

The result from management's point of view is that as they become more rationalized and efficient, these robotized factories will come to resemble continuous-process plants such as oil refineries more nearly than they will conventional manufacturing organizations. For example, the number of workers who are supervised by a manager will tend to increase, as will the corresponding ratio of managers to workers. Individual discretion in making work-related decisions will decline—we will be more interested in whether an employee knows the "right" solution to a problem than in whether that employee can exercise good judgment and creativity.

CHANGING WORK ROLES FOR HUMAN LABOR

Combining Technologies in Organizations

The effects of the technological revolution that has been described will be more pronounced but, I think, less severe than most popular writers have imagined. The next chapter describes the effects that may be experienced throughout society within the next 10 years. For the present, I will discuss only the changes that robots are likely to cause within manufacturing organizations.

I have described organizations as if each employed a single technology. Although a single technology may typify an organization, however, organizations generally employ many. For example, a factory that has been designed for mass production typically will be part of an organization that includes an R & D unit, a sales force, a personnel department, and so on. The technologies used in each part of this organization are different.

Even a computer-controlled factory will include units that operate diverse technologies, although, for the reasons that are discussed above, such diversity will be reduced relative to that found in conventional factories. Nonetheless, the presence of such diversity constitutes a source of problems for management.

Coordination of Technologies

It always is necessary to coordinate the efforts of different organizational units. In the production planning illustration, such coordination is seen to take two alternative forms. In one case, the plans of each unit are integrated with the plans of other units informally. Experienced managers and technicians discover that one unit, in doing what seems best for that unit, has begun to operate in a way that creates problems for other units. Tradeoffs are proposed between the one unit and the others. Negotiations ensue. Frequently, conflict erupts among the units in question. Ultimately, such differences are resolved, and a workable resolution of the differences between the units—for example between marketing and production—is adopted. Most often, however, the tradeoffs that result are less than optimal from the organization's point of view.

The other case is one in which a well-structured technology can be used to plan the activities of all related units, as is the case when the activities of units in a manufacturing firm are coordinated by a computer-based planning system. In this case, the units do not plan for themselves; rather, they supply information to the planning system, which is, in a sense, above the operating units in the organization's hierarchy of authority.

It is the latter type of planning system that will be used in conjunction with robots. Nonetheless, no planning system can incorporate effectively all aspects of organizational functioning. Even in the most technologically advanced organizations, there will remain units that must employ ill-structured technologies, simply because we do not fully understand everything an organization does, and have not been able to reduce all organizational activity to a science. Thus, some parts of every organization must rely on judgment in the absence of a well-structured technology.

The management problem to which I referred earlier is to coordinate the activities of these units with the activities of the remainder of the organization, which in turn are coordinated by a well-structured technology. Skillful management is required to blend judgment and intuition on the one hand with science on the other in the interest of good organizational performance. Organizations that have been successful in doing this routinely have employed liaisons to mediate between one part of the organization and the other. Individuals who have worked in both parts of the organization—who consequently have come to know how each part must function—typically are chosen for the task. Because their activities are essential to the smooth functioning of the organization, these individuals typically enjoy high status as intra-organizational liaisons.[4]

The role of liaison certainly is not new to manufacturing organizations; liaisons typically have mediated differences between marketing and engineering, between production units and major customers, and between R & D and engineering. However, as the technologies employed by organizations become more diverse—that is, as production systems move toward the lower left-hand corner of Figure 6–1 while R & D remains in the upper right-hand corner—the importance of the role to the organization will increase.

Occupational Diversity

Overall the variety of occupations employed in manufacturing will be reduced, relative to conventional manufacturing operations, by the replacement of many ill-structured technologies by well-structured ones. The diversity of skills, aptitudes, education, and wage rates found in these organizations will decline accordingly. . Skilled and semiskilled workers will be replaced by machines, leaving highly trained technicians and engineers and such unskilled workers as janitors, parts handlers, and the like to populate the fac-

[4]Paul R. Lawrence and Jay W. Lorsch, *Organization and Environment: Managing Differentiation and Integration*. Boston: Division of Research, Harvard University Graduate School of Business Administration, 1967.

tory floor. If you are inclined to classify jobs as being good or poor according to the amount of skill, education, intelligence, and discretion that they require and, in addition, by how much they pay, then you will conclude that good jobs will be lost at the expense of poor ones. The best jobs—those of managers, engineers, and technicians—and the poorest ones will remain after robots have taken over many, if not all, of the better blue collar jobs. Thus, to a greater extent than ever before, factories will employ two distinctly different workforces. Techniques for managing, compensating, and in general motivating the efforts of one group of employees probably will not be effective with the other. I have described the management techniques that are appropriate for workers in either category in a recent book, which the interested reader may wish to pursue.[5]

In any event, the problem of recruiting and managing the unskilled segment of the factory work force probably will be far less vexing than the problem of maintaining a cadre of technicians and engineers. As I noted earlier, our secondary and post-secondary school systems are deficient in terms of providing the training in science and mathematics that is required by conventional industry. Our needs for such training will increase several-fold as production technologies become more complex. One resort, the one to which industry undoubtedly will turn, is for employers to shoulder an increasing share of the burden of educating their workforces. This will become inevitable in the long run as accelerating technological change shrinks the useful life of a secondary and post-secondary education. In the short run, I am afraid that employers will also be forced to provide what is essentially remedial education in basic scientific, mathematical, and communication skills.

Nonetheless, in the longer run, society, including our educational systems, must restructure itself in response to pressures of the new industrial revolution. Western society has changed many times in centuries past as science and technologies have changed. Unlike earlier changes that were, in a sense, evolutionary—gradual, for the most part, and not planned—future changes will be more abrupt, discontinuous, and intentional. The reasons that lead me to believe this to be the case are the topic of the concluding chapter.

[5]Robert A. Ullrich, *Motivation Methods That Work*. Englewood Cliffs, NJ: Prentice-Hall, 1981.

7

The Next Industrial Revolution

WHAT IT WILL MEAN
TO INCREASE PRODUCTIVITY

A Changing Rate of Change

For most of us, one day still seems to be pretty much like the next. Looking across generations, though, it becomes clear that each generation matures in a world that did not exist when its members were born. My parents, for example, were alive at the time the Wright brothers flew. Since that time, their generation has seen a robot land on Mars. The number of major historical events that have occurred between these two events is incredible. More incredible, but less obvious, is the increasing frequency with which events of that order of importance are occurring.

The authors of our Constitution saw the need for universal literacy in the government of the people that they created. Thus, society shouldered the burden and expense of providing free, compulsory education. We had then the luxury of a new and slowly changing nation with which to work: there remained generations within which the goal of universal literacy could be pursued. That luxury is lost—forever, it would seem—because everything that was learned in the 200 years that followed paved the way for learning even more. Each discovery made possible more discoveries, and knowledge snowballed, rendered changes in society, and pointed the way to further discoveries.

We have outlived the time when society could adapt gradually, through trial and error, to new ideas. Now, for example, new industries and occupations appear within the span of a single generation while others disappear through obsolescence. Two consequences of the accelerated rate of change are of vital importance to our view of the future: one is that, unlike our elders, we are in a position to foresee and anticipate with some accuracy what will happen next. Many of the major changes in technology that will occur within the next 10 years can be seen today in our R & D laboratories. The other consequence is that it is now incumbent upon us to anticipate—that is, to plan—the ways in which such changes will, or will not, be used and the alterations of society that will be needed to support such changes if we choose to implement them. Much less of the future will merely happen to us. Much of what will happen in the next 10 to 20 years will be a result of what we choose to do today. We can choose to develop and employ a robotics industry, or we can sit back and watch other nations do so.

Robotics and Unemployment

Some of the people to whom I have spoken about robots conclude that businesses, in their relentless quest for profit, ultimately will consign hundreds of thousands of employees to the welfare rolls and replace them with robotic machines. Such thoughts are upsetting to decent people, who want to place the well-being of their neighbors above the quest for efficiency and profitability. Their conclusion seems to be a logical, although unpleasant, one to draw when one reads that as much as 75 percent of the blue collar workforce in manufacturing eventually will be replaced by robots.

I think that this conclusion, far from being logical, is ill-founded. For one thing, the rate at which the workforce is growing will decline rapidly over the next few years. Between 1966 and 1976, young adults born during the baby boom, and women, who previously had not sought careers, entered the workforce in such numbers that they increased it by nearly 50 percent.[5]

On one hand, this suggests that our economy has the capacity to absorb great numbers of workers when it is functioning properly. Thus, it is not necessary to conclude that workers who become displaced by robots will be unable to re-enter the workforce. On the other hand, demographic data lead one to conclude that the next 10 years will see an absolute shortage of people entering the workforce. There are fewer young people to employ now that the baby boom generation has gone to work, and the majority of women who want to be permanently employed probably are employed today.

The inevitable growth that will accompany economic recovery could be confronted by a relative scarcity of new workers. Ironic as this may seem with upward of 10 million workers unemployed in 1983, it seems reasonable to conclude that robots have appeared just in time to avert a labor shortage that otherwise would rekindle the fires of inflation.

There are still other reasons to be optimistic about the long-term effects of a robot workforce. Many of the dire projections of mass unemployment are based on the assumption that robots will compete with people in nearly every line of work. The scenarios from which these projections are drawn have robots working at con-

---

[5]The baby boom is attributed to the reunion of young men and women who had been separated from one another during World War II. If this was, indeed, the cause of the baby boom, I would like to add that it was one hell of a reunion—the baby boom lasted close to 20 years.

struction jobs, doing household chores, waiting on tables, and even, as mentioned in Chapter 5, conducting psychiatric interviews. There is little reason to believe that employing robots at such tasks as these will be feasible, not to say economical, five, 10, or even 20 years from now. The basic elements of technology that will be needed before robots can perform such tasks are not on the horizon, as far as I can tell.

Other scenarios portray the time when all positions in the productive sector of the economy will be held by robots, while humans work at service tasks. The question that this scenario has raised for at least one wag is, "Will machines produce goods while humans are reduced to working as hairdressers?" Actually, after growing at a greater rate than the productive sector for many years, the service sector of our economy now employs a greater number of people than are employed in productive industries. That does not mean, however, that we are becoming a nation of barbers. An example will be instructive. The petroleum industry employs people by the hundreds of thousands, yet their refineries and chemical plants are capital intensive, requiring relatively little human labor. The majority of workers in this industry are employed in marketing, transportation, data processing, research, and other service occupations. The automation of productive processes in this industry coincided with growth in the number of people it employs and, perhaps, with improvements in the quality of the worklife that they enjoy.

Quality of Life

Nonetheless, one current trend casts a melancholy pall on the vision of the future that is emerging here. If we take wage rates as measures of the quality of jobs, then it appears that growth in the service sector of the economy has created more poor jobs than good ones. Educators, physicians, and musicians are employed in service occupations, but so are retail clerks, car wash attendants, and garbage collectors. The latter jobs are poor—they pay little and lead nowhere. More poor jobs are created in the service sector than good ones. Moreover, these jobs do not compare favorably even with jobs in manufacturing that require comparable levels of education and skill but provide better wages—the kinds of jobs that robots will take over. As this trend continues, the gulf that separates low- and middle-class Americans will become wider. The problem of a highly stratified work force that I foresee in manufacturing organizations probably will parallel a similar, but greater, problem in society unless society is prepared to change.

WORK, LEISURE, AND ROBOTS

Labor Productivity
Vs. the Productivity of Capital

Every school child knows that in the late 1800s more than nine out of 10 Americans were living on farms. The average work week then was about 65 hours a week. If you have ever lived on a farm, you know that such notions as *fixed* and *finished* are fictional concepts, devised by city dwellers and having no counterpart in reality. Work is never done on a farm—there is little time for leisure.

With the industrialization of America, we literally emptied our farms into the cities—fewer than four Americans in 100 work on farms today. Mechanization permitted a dramatic increase in farm productivity along with a drastic reduction in the amount of labor required to sustain that productivity. A similar trend was occurring in our factories as mechanization increased the output of each worker. So it seemed, anyway.

I do not believe that a worker's productivity increases per se when the technology of work is improved. The field hand who tilled the fields with a horse and plow could till more land in a day after the farm owner purchased a tractor for him to use. However, although it is conventional to say that the farm hand's productivity increased when he began to drive a tractor, I am led to conclude that the productivity of the farmer's capital increased when he invested in the tractor, whereas the farm hand's productivity remained about the same as it had been. It is worth noting that wages paid to farm labor have remained among the lowest paid to any occupational group. Apparently, farmers recognize the difference between returns to capital and labor productivity.

If you have lived on a farm, you also know that a good farm hand is a competent mechanic, carpenter, and lay veterinarian as well as a skilled operator of agricultural machinery. Yet, a relatively unskilled assembly line worker can earn a better wage. The reason for this is that the assembly line worker, unlike the farm hand, was able to unionize and, consequently, to share with the factory owners the increasing returns to capital investment. The argument that such capital investments increased the workers' productivity and, thus, their entitlement to higher wages was a rationalization that made more agreeable what otherwise was a political division of wealth between capital and labor.

There are many ways in which such wealth can be distributed to workers. The two of immediate interest are increased wages and increased leisure. Until World War II, labor chose to pursue both, and as wages increased, the work week decreased to about 40 hours. From World War II to the present, labor has shown little interest in further reducing the work week: virtually all gains have been taken in the form of increased wages and fringe benefits.

That observation has a chilling effect on the speculation that robots, as they increase factory productivity and reduce the need for human labor, will permit a shorter work week and more leisure-time pursuits. Americans, faced with a choice of higher wages or shorter hours of work, elect the higher wages: they do not want to trade an increased standard of living for more leisure time. For one thing, aspirations seem to have changed. We want more today rather than just more time in which to enjoy what we presently have. For another, I am not sure that the average person has the capacity to enjoy and, consequently, to benefit from greater leisure.

There was a time when utopian thinkers were convinced that freedom from toil would permit a flowering of civilization, that once freed from drudgery, humanity would devote itself to finer things— to education, the arts, science, and discovery. As it seems to have turned out, humanity elected to devote itself to watching bowl games on TV.

For all its faults, a place of work provides a community of sorts where interpersonal bonds are formed to take the place of tribal or village life. Interesting things happen at work. There are power struggles and scandals to enjoy, games to play with management, and interoffice rivalries to be fought. In most respects, work is more interesting than cutting the lawn or watching TV. Little wonder, then, that we seem to have had our fill of leisure time.

To the observation that Americans seem to have their fill of leisure, though, I must add another. Essentially, there are two categories of reasons that explain the robot's recent emergence as a competitor with human labor. First, robotic and, particularly, microcomputer technologies have advanced to make such competition feasible. Second, labor costs have grown at a much greater rate than has the cost of robots. The matter of relative cost puts labor in a bind. The higher hourly wages go—either as a result of gaining a shorter work week while keeping take home pay constant or as a result of maintaining the present work week and demanding greater wages—the more attractive robots will become as substitutes for human labor. The alternatives to labor are two: to compete with robots by working for lower wages or to move into occupations that cannot be performed by machines.

My guess is that both will occur. First, the threat of competition from robots will temper and depress labor's demands for future wage increases in manufacturing industries. Second, a shift in employment will follow that is similar to, but less drastic than, the shift from agricultural to factory employment that occurred at the beginning of the century. For a shift of this sort to occur, our present educational system must be changed.

CHALLENGES TO EDUCATION

Education for Work

I have had three full-time occupations in my life so far. As a merchant seaman and later as an educator, I followed ancient callings. In between, I worked with computers in an industry that did not exist when I was born. Other industries have developed since that time: television, antibiotic drugs, microcomputers, genetic engineering, robotics, plastics, and nuclear power, to name a few. Our future lies not in competing with machines that can perform simple, repetitive tasks, but in discovering and developing new industries and more profitable things to do. We should not hesitate to replace laborers with robots any more than we hesitated in substituting earth-moving equipment for ditch diggers.

Society faces a challenge today that it has not faced before. For example, ditch digging as an occupation depended on laborers, shovel manufacturers, and supervisors. None of these occupations required much in the way of intellect or education. To replace the ditch digger with a machine, however, required engineers to design and produce it, skilled mechanics to keep it running, a petroleum industry to power it, and, not the least, a skilled heavy equipment operator to make it work. Each occupation required much more education, skill, and experience than the occupation it replaced. Relative to bulldozers and backhoes, microcomputers and robots are far more complex and more difficult to design, build, and operate successfully. Considerable education is demanded of those who will use them. Who will provide this education?

The view from inside academia is not encouraging. The question of whether public primary and secondary schools in general are functioning adequately is beyond the point of being subject to debate. Declining Scholastic Aptitude Test scores highlight serious problems in our schools. I think it is fair to say that, on balance, schools provided better education 20 years ago than they do today, and that the decline set in when socialization replaced education as the schools' primary goal. It was an entirely decent but badly misplaced sense of fairness that led us to burden the schools with the task of righting society's wrongs. It is hard enough to teach people to think and write clearly; it is nearly impossible to do this while taking on the responsibilities that families and communities seemed all too willing to relinquish. The result has been that academic standards have been dropped so low that the near-illiterate can attain them while the gifted fail to realize their intellectual potentials. Colleges, for their part, dropped their own standards accordingly.

The other major problem is that the curricula offered by our colleges have not kept pace with the times. Liberal arts majors, for example, can successfully avoid coursework in mathematics and English composition as they meet the requirements of their degrees. How will they work if they are mathematically illiterate and cannot write adequately? It is the least of my intentions to suggest that liberal education be made vocational. I merely want to suggest that a liberal education should not permit such fundamental areas of ignorance.

American business has gone soft and is losing ground to competitors, such as the Japanese, who are willing to work a little harder with a little less. Our educational enterprise has gone even softer and lacks the fortitude to demand excellence.

The United States no longer has the world's highest per capita income; we do not have the world's lowest infant mortality rate; we are far from having the lowest crime or illiterary rates. Nonetheless, we are the leader in scientific and technological innovation, in Olympic athletics, and in space exploration. We are, in a sense, a nation of élites. The élite—the champions who rise above their peers—are few, however. Unless our school systems rise to the challenge of equipping the majority of the population to compete in the coming industrial revolution, only the privileged few will emerge as champions. The vast majority will be too ill-equipped to share the responsibilities as benefits of the future.

I do not mean that the vast majority of Americans will subsist in poverty, but they will find themselves increasingly ill-prepared to understand and control the society in which they live. Work, from agriculture to manufacturing, is becoming increasingly complex. The technologies employed will become increasingly well-structured, but also, as a result, more complicated. One must either be prepared to understand this complexity or be willing to accept the consequences of not understanding it. Only those whose education has prepared them to understand it will be able to control and, thus, benefit from its use.

Today's robot essentially is made to order; tomorrow's will be mass produced by other robots. In consequence, the cost of robot labor will decline. One estimate suggests that today's $40,000 unit will cost about $3000 within 10 years or so.[6] By that time, factories and warehouses that operate virtually without blue collar workers will be commonplace. Hourly wages will not be paid in these factories; the benefits of operating them will accrue to their owners, to managers, and to technicians. People who do not understand how such factories work will become "none-of-the-above."

The robot of which I have written is only a symbol for the new technology that is developing. Computers and computer-controlled equipment of many kinds will relieve us of our toil, but also of our ability to influence and benefit from the organizations in which we are employed. I see only two profitable ways to get by in the future, and each requires at the least a rigorous, demanding education. One is to know the new technology well enough to be able to control it—that is, to design it and make it work. The other is to know things that computers can never learn.

[6]James S. Albus, *Brains, Behavior, and Robotics*. Peterborough, N.H.: BYTE Books, 1981.

Bibliography

The following annotated bibliography includes most of the titles on robotics that are listed in the 1982 edition of *Books in Print*. Few in number, these books represent the bulk of the published literature in the field, aside from articles in scientific journals and in the popular press. The brevity of this list persuaded me to write the present book.

Initially, I was curious about robots and merely wanted to learn more about them. What I was looking for at the time was a book on the order of this one. Unable to find one, I purchased most of the available literature, diverse as it is, and pursued journal articles, news stories, and personal interviews with people in the field. One important source of information that I found was the promotional literature that robot manufacturers graciously gave me for the asking.

As I studied the materials that I had collected, I began to appreciate the need for a book that would provide an introduction to the field for persons who were interested, as I was, but had no prior knowledge of robotics. And so, I wrote the book.

I have arranged the following bibliography for the sake of convenience, by categorizing the books according to the interests that they address. The annotations with a few exceptions were written by a graduate assistant, Jim Woodrow.

BOOKS

ENGELBERGER, JOSEPH F., *Robotics in Practice*. New York: AMACOM, 1980. 291 pp., $39.95.

Focused on robots in the factory, *Robotics in Practice* is an attempt to "... allay ill-founded concern (about this new technology) and to eliminate some of the pain that inevitably accompanies the adoption of unfamiliar concepts." Divided into two parts, *Robotics in Practice* progresses from Part I on the fundamentals and management of robots to Part II on factory applications of robots. In Part I, major emphasis is placed on nine topics, that include: the use of robots in manufacturing; robots' anatomy; end effectors (grippers, hands, pickups, and tools); matching robots to the work setting; factors related to their maintenance, reliability, and safety; establishing plans to implement them; and the economics, sociological repercussions, and future capabilities of industrial robots. Part II describes present applications of robots to more than a dozen industrial processes beginning with die casting and ending with the manufacturing of glass.

Illustrations and color plates provide the reader with numerous examples, making concepts and applications clear and simple to understand. The appendix provides a directory, purportedly not exhaustive, of principal robot manufacturers in Europe, the United States, and Japan. In summary, the book presents a comprehensive, although somewhat dated, introduction to industrial robots.

SUSNJARA, KEN, *A Managers Guide to Industrial Robots*. Shaker Heights, OH: Corinthian Press, 1982. 181 pp. no price given.

This book takes up where Engelberger's *Robotics in Practice* leaves off. Written for managers with little prior experience in robotics, Susnjara's book contains a good deal of practical advice and suggestions.

Two initial chapters are devoted to a description of the technology, but better, more complete descriptions have been provided by other authors. Chapters three through six contain this book's best contributions—advice from the firing line on what to look out for and what can go wrong, procedures to determine where robots can be used to advantage and where they can't, and a list of simple rules that can guide the manager throughout the robot's installation. Chapter seven anticipates the future of robots in automatic factories. The next chapter provides some generally useful advice for managing employee relations in plants where robots are introduced. Chapter nine includes a straight-forward, but rather simple series of calculations for providing the economic justification of an investment in robots. Chapter ten takes the obligatory look at the future.

The first appendix is a workbook and installation manual for robot applications. Appendix II contains a glossary of terms. On balance, this book contains much that should be of interest to a newcomer to the field of robotics.

TECHNICAL DATABASE CORPORATION, *The Robotics Industry Directory*. La-Canada, CA: Technical Database Corp., 1982. 238 pp. $37.00 (paperbound).

With a monthly update, this annual directory is the ultimate easy-reference directory of products and firms in the robotics industry. The directory includes 12 sections, the first of which contains four matrix comparisons of industrial robots in terms of their applications, sensors, power drives, and price ranges. The next section, consisting of more than 100 pages, is a comprehensive listing of manufacturers of industrial robots, with approximate costs, general equipment specifications, delivery times, lengths of warranties, cities (worldwide) with marketing or service offices, and the names and addresses of the manufacturers' sales representatives.

Other sections include listings of British robot manufacturers, European robot associations, industrial robots manufactured in Japan, special application systems, CAM systems, distributors, and components and peripherals. Two important sections in the back of this directory contain the names and addresses of consultants and research institutes in the field of industrial robotics. The glossary defines over 60 terms used throughout the *Robotics Industry Directory*. Finally, the index is divided into sections that correlate with the sections located in the table of contents.

KENT, ERNEST W., *The Brains of Men and Machines*. New York: McGraw-Hill, 1981. 286 pp. $8.95 (paperbound).

Once it was believed that we could learn to build better computers by learning how the human brain functions. It turns out that, to date, the opposite has been true; we have learned something about the brain from our knowledge of computers.

Kent uses elementary computer technology to explain how the human nervous system may function, and, in doing so, provides a clear, but elementary, description of how computer technology can be applied to such human tasks as robots may perform.

Although the book assumes some basic knowledge of computer circuitry, the reader who lacks such background can learn enough from Kent's lucid examples of that technology to follow his entire argument.

WEIZENBAUM, JOSEPH, *Computer Power and Human Reason*. San Francisco: W. H. Freeman and Co., 1976. 300 pp. $9.95 (paperbound).

Weizenbaum, whose work is introduced earlier in this book, has produced an essay that attempts to place the computer within its proper context in our society. The focal point of the essay is an analysis of what computers cannot and should not do: an intelligence that cannot share the human experience should not be used in every instance that it might be used.

To bring the reader to his conclusion, Weizenbaum introduces the notion of tools and then describes the computer as a universal tool. The scientific enterprise is introduced next, with its proclivity to become enamored of such tools and, consequently, to become blind to the essence of what it studies. In this manner, the limitations of science, computers, and artificial intelligence are revealed to the reader.

SMITH, DONALD N. AND WILSON, RICHARD C., *Industrial Robots a Delphi Forecast of Markets and Technology*. Dearborn, MI: Society of Manufacturing Engineers, 1982. 86 pp. no price given

The Delphi forecasting technique was developed to explore the consensus of opinion on future trends of selected groups of experts. Typically, a questionnaire is sent to members of a Delphi panel, who have been recruited to participate in the forecast in question on the basis of their experience and expertise. Questionnaire results are summarized, and the resulting data together with written comments that frequently are volunteered by panelists are distributed to the panel, along with a second questionnaire. Panelists, in responding to the second questionnaire, may choose to modify their responses in light of the responses to earlier questionnaire items provided by other panel members. Occasionally, a third round of questions is used, along with summarized results from the second questionnaire.

Smith and Wilson, both of the University of Michigan, have conducted a Delphi forecast for the robotics industry using a panel of 100 individuals employed in organizations that supply or use this equipment. The data that are published in the present volume are an interim summary of the forecast's results to date.

Forecast results are aggregated under three major topic headings: the technological, marketing, and sociological implications for this industry. One hundred and thirty-eight pages of appendices report the detailed results of the survey, describe the Delphi technique, provide a glossary of terms used in the survey, and list the tables that appear in this report.

ROBOTICS INTERNATIONAL, SOCIETY OF MANUFACTURING ENGINEERS, *Robotics Today '82 Annual Edition*. Dearborn, MI.: Society of Manufacturing Engineers, 1982. 369 pp. no price given.

The '82 Annual Edition is a collection of articles that were published in *Robotics Today* between summer, 1979 and fall, 1981. Chapter one comprises twenty-two articles under the heading, basics. The second chapter, on human factors, includes only three articles, perhaps illustrating a dearth of knowledge in this area. The following six chapters are devoted to specific applications: machine loading and material handling, welding, assembly, casting and forging, painting, and other applications. Chapter nine comprises three articles on the implementation of robots, and chapter ten contains nine articles on vision and sensing.

An index to *Robotics Today* proceeds a topic index to the '82 Annual Edition. Following is a directory of manufacturers and distributors that includes an alphabetical listing of the major robot suppliers and basic specifications of the equipment handled by the most active suppliers. The volume concludes with an annotated bibliography of technical papers available from RI/SME and a topical index to these papers.

ALBUS, JAMES S., *Brains, Behavior, and Robotics.* Peterborough, NH: BYTE Books, 1981. 352 pp. $16.95.

Written in the style of a college text, this book is divided into three major themes. The first four chapters are devoted to a description of the human nervous system. Chapters 5 through 7 discuss approaches to modeling the neurological system and the higher functions of the brain. The final five chapters discuss robots, control systems, artificial intelligence, and their applications and implications for the future.

McCORDUCK, PAMELA, *Machines Who Think.* San Francisco: W.H. Freeman and Company, 1979. 375 pp. $8.95 (paperbound).

Intended for the nonspecialist reader, *Machines Who Think* is largely a history of the artificial intelligence (AI) movement. The author focuses on this topic because ". . . artificial intelligence in one form or another has pervaded Western intellectual history." A major portion of this book is devoted to interviews with 36 prominent figures in the artificial intelligence movement, who generally are credited with having transformed this art into a fledgling science. One reason for using interviews, states the author, is to convince skeptical humanists that science is not alien to the humanities.

In the introductory chapter of this book, the reader journeys back to mythic times when attempts to create artificial intelligences were purported to have begun. The chapter also contains a brief summary of "Puritan distinctions" among the three basic artificial intelligence machines: automata, androids, and robots. Chapter 10 is devoted exclusively to robotics and general intelligence. The last chapter assesses the use of symbols embodied in a physical system, describes the function of artificial intelligence as a conduit between science and art, and looks at potential roles of artificial intelligence machines in the future. Quotations and illustrations throughout this book illuminate both the history and the possibilities of the movement.

JOURNALS

Robotics Age is published six times a year by Robotics Age Inc., Strand Building, 174 Concord St., Peterborough, NH 03458. $15.00 per year.

Robotics Today is published quarterly by Robotics International of the Society of Manufacturing Engineering, One SME Drive, P.O. Box 930, Dearborn, MI 48128. $24.00 per year.

TECHNICAL PAPERS AND CONFERENCE PROCEEDINGS

TANNER, WILLIAM R. (Editor), *Industrial Robots* (2 volumes) Dearborn, MI: Society of Manufacturing Engineers 1979. 574 pp. $24.95 per volume.

Published in two volumes, *Industrial Robots* is a compilation of articles suitable for persons interested in an in-depth study of robots in the factory setting. Volume I focuses on fundamentals and capabilities—robot automation, definitions of some basic terms, the cost of robots, their potential economic return and safety, the future of robotics, and a description of some present practical applications of industrial robots. This volume is made up of 28 articles, a list of manufacturers and distributors, and a list of robot organizations. Volume II, a collection of technical papers, application notes, and trade journal articles, contains 12 chapters.

Although the papers in the two volumes have become dated in light of the rapid advances made in this field, they provide basic information.

BROCK, T. E. (Editor), *Proceedings: 3rd Conference on Industrial Robot Technology and 6th International Symposium on Industrial Robots.* Oxford, England: Cotswold Press Ltd., 1978. $79.50 (paperbound).

This selection of technical readings on industrial robots offers professionals in the field a broad sample of scholarly statements and opinions on the state of the robotics movement. Sir Ronald McIntosh, Director General of the National Economics Development Office of the United Kingdom, provides the opening remarks. The table of contents lists the titles of 39 papers presented in this volume. Contributors to these papers represent 10 European countries, the United States, Japan, and the Soviet Union. Numerous diagrams and illustrations appear throughout the papers. Comprehensive lists of the delegates in attendance at these proceedings are categorized by individual name, organization, and country and are located in the concluding pages of this volume.

· FOR THE HOME ROBOT BUILDER

HEISERMAN, DAVID L., *Build Your Own Working Robot*. Blue Ridge Summit, PA: TAB Books, 1976. 234 pp. $8.95.

Written for hobbyests, this manual provides step-by-step instructions for building a robot. The author suggests several requisites before undertaking the robot's construction: first, have a background in basic electronics; second, possess a working understanding of Boolean algebra; and third, have a mastery of the fundamental hands-on skills of electronics.

The robot's construction can occur in three separate stages. All three stages represent different phases of the robot's development. The completion of each stage will provide the builder with an operational robot that can be enjoyed until the builder is ready to proceed to the next stage. Numerous diagrams supplement the readings.

LOOFBOURROW, TOD, *How to Build a Computer-Controlled Robot*. Rochelle Park, NJ: Hayden Book Company, Inc., 1978. 132 pp. $9.75 (paperbound).

This "how to" book sets forth directions for building a computer-controlled robot. Persons interested in a hands-on experience with robotics will want to rely on this book as a reference manual. *How to Build a Computer-Controlled Robot* provides the reader with a basic education in robotics and microprocessors. Readers, suggests the author, should have a basic understanding of electronic circuit diagrams and possess a reasonable degree of mechanical ability before undertaking this robot-building project. Diagrams, photographs, and tables provide additional directions for the robot's construction. When built, this robot has the capacity to function independently or it can be operated manually with joystick.

HEFLEY, ROBERT M., *Robots*. New York: Starlog Press, 1979. 98 pp. $6.95 (paperbound).

The author, a professional actor and theatrical director, published this photo guidebook as a review of the various roles that robots have performed in film and on television. The opening pages summarize the literary classics that have served as models for Hollywood. Before the word *robot* was even coined, Hefley claims, Hollywood was showing these mechanical-looking creatures in silent films.

The book is divided into four main sections. The first provides a chronology of robots in films from 1897 to 1979. The second section is in color, and includes photos from such films as *Star Wars* and *Forbidden Planet*. The hird section, also in color, presents photos of robots that starred in such television shows such as *Star Trek*, *Frankenstein Jr. & the Impossibles*, and *Battlestar Galactica*. The last section contains a chronological review of television robots from 1940 to the 1970s. This photo guidebook is suitable for most ages and all levels of interest.

GEDULD, HARRY M. & GOTTESMAN, RONALD (Editors), *Robots Robots Robots*. Boston: New York Graphic Society, 1978. 246 pp. $14.95.

The selections included in this book range from "Maelzel's Chess-Player" (1836) by Edgar Allan Poe to "Silent Sam" (1969), and from "Robots in the Nursery" (1958) to excerpts from *The Making of Frankenstein's Monster*, a novel published in 1838. Four main sections include over 20 readings. Complementing the readings is a series of drawings and photographs of machine art and an array of scenes from robot movies. An extensive bibliography lists references to robots in poetry, fiction and nonfiction, and drama and technical studies. The concluding pages provide a chronological outline of robots in the cinema from 1897 to 1977.

CHILDRENS' LITERATURE

BENDICK, JEANNE, *Super People: Who Will They Be?* New York: McGraw-Hill Book Company, 1980. 128 pp. $7.95.

Illustrated with old prints, line drawings, and photographs, *Super People* examines clones, bionic people, robots, and other contemporary topics. Ostensibly published for children, this book may appeal to readers of any age, who will enjoy its creative style. A 32-page section on robots concentrates at an elementary level on the evolution of robots, on their anatomy and their capabilities, and on several robot celebrities. A clever "score card" in the last pages of the book identifies strengths and weaknesses of robots and other "super people."

KLEINER, ART, *Robots*. Milwaukee: Raintree Publishers, 1981. 48 pp. $10.25.

Written for children, *Robots* describes how a robot operates and reviews the history and development, utility, and future of these intriguing machines. Illustrations help the young reader to visualize new concepts. The glossary provides an elementary definition of robotics terms used in the book. A professor of mechanical engineering consulted with the author—adding validity and integrity to the book. Colorful and succinct, adults will enjoy perusing this book too.

OTHER BOOKS

CHEN, WAYNE, *The Year of the Robot*. Beaverton, OR: Dilithium Press, 1981. 182 pp. $7.95.

The author, dean of the engineering school at the University of Florida, provides the reader with a look into the future, when artificial intelligence will perform important functions in everyday life. Part I is a philosophical treatise, while Part II is a novella about love and conflict. The novella is the author's attempt to ". . . bring to the reader one of the great scientific creations"—the robot. Chapter 6 of Part I contains mathematical charts and formulas that ". . . establish the technical validity of the intellectual robot." The reader can choose to read this chapter or omit it, without affecting the underlying theme of the second part. The concluding pages of this book consist of two complex appendices.

SAFFORD, EDWARD L., Jr., *The Complete Handbook of Robotics*. Blue Ridge Summit, PA: TAB Books Inc., 1978. 358 pp. $14.95.

In this handbook's preface, the author states: "This *is* the dawning age of robotics." But what is robotics? The term *robotics* simply means robot technology, which suggests the research and development of ". . . machines able to do those things *we* don't want to or cannot do, but which must be done somehow to make our life more pleasant and satisfying."

Divided into 14 chapters, this handbook discusses a broad array of topics on robotics. Chapter 1 defines for the uninitiated such terms as cybert, machine-automated, cyborg and cybot (two counterparts), android, robot, and humanoid. The chapter continues to discuss three basic questions: why robots, are they trustworthy, and what can they do? The chapter concludes with such diverse topics as: creating a robot, the Robot Master, human fears of robots, and some general capabilities of robots. Chapters 2 through 12 provide technical discussions of robot mobility, sensors, primary power sources, the basics of robot systems, servomechanism systems, the robot brain, commercial robots, the construction of a hobby robot, interfacing robots with computers, and radio control of a robot. The first chapter and the last two chapters will contribute to the non-specialist reader's needs. The last two chapters provide a general discussion on modern robots, robot societies (i.e., industrial and hobby organizations), and the future of robots.

WINKLESS, NELS & BROWNING, IBEN, *Robots on Your Doorstep*. Portland, OR: Robotics Press, 1978. 179 pp. $6.95.

This book about thinking machines is written by a founder of the American Robotics Society and a robotics pioneer. The 10 chapters that comprise the book will, according to the authors' introductory remarks, ". . . amaze you, frighten you, nauseate you, excite you, enrage you, disgust you, and shock you."

Index

A

Albus, James S., 104, 111
American Robotics Society, 117
Artificial intelligence: *See also* Problem
    solving; SHRDLU
  Doctor, example, 72–73
  Parry, schizophrenia model, 73
  and psychological testing, 71
  Turing test, 72
Automation, early, 11

B

Babbage, Charles, 11
Bendick, Jeanne, 115
Brains, Behavior, and Robotics, 104, 111
Brains of Men and Machines, The, 60,
    109
Brock, T.E., 113
Browning, Iben, 116
Bublick, Timothy, 45
Build Your Own Working Robot, 76, 113
Bureaucracy, and robotics:
  introduction of robots, 90
  and production planning, 90
  and structures of tasks, 89
Bureaucracy, in organizations:
  planning in, 87, 88
  pricing in, 88
Business Week, 54

C

CAD/CAM, 52
Calculator cases, molding by robots:
  closure of dies, 31
  defects, sensing of, 30, 31, 32
  dies, 29–30, 32
    changing of, 32
  grasp, sensing of, 33

hands, number of on robot, 33
hands, removal of from work area, 31
inserts, in molded parts, 32–33
overgrasp, 30
packing of parts, 33–34
part removal, 30, 31
quality control, 33
runners, 29
Capek, Karel, 39
Capital invested per hour of labor, 5
Charles Strak Draper Laboratory, 59
Chen, Wayne, 115
Complete Handbook of Robotics, The, 116
Computer Power and Human Reason, 74,
    109–10
Computers, progress of, 35
Conigliaro, Laura, 50
Continuous path robots, teaching of, 21

D

Datamation, 5
Democratic organizations, 88–89
  effectiveness, concept of in, 88
  feedback in, 89
  and solutions for ill-structured prob-
    lems, 88
Diversity of occupations, 92–93
Dizard, John W., 41
Dunnavant, Bob, 7

E

Education, challenges to:
  author, experience of, 102
  class conflicts, 103, 104
  ditch digging, paradigm of increasing
    complexity, 102–03
  mathematics, illiteracy in, 103
  robots, drops in costs of, 104
  schools, loss of standards in, 103
Employment, increase of in U.S., 5
Engelberger, Joseph F., 55, 107
Environments, harsh, and robots, 41–42
Evolution, problem of, 81

Technology:
  combinations of, 90–91
  coordination of, 91–92
    liaisons, in organizations, 92
    tradeoffs, 91
    and types of problem structures, 91–92
  organization of:
    and computers, communication with, 83–84
    crafts, 87
    hardware, 87
    ill-structured, 84
    planning, 84–85
    R and D, 86
    tasks, structures of, 85
    types, 86
    well-structured, 84
  rapidity of advance of, 97
Technology Review, 53
Tennessee Fan Company, 3, 4
Time, 6
Turing, Alan, 72

U

Ullrich, Robert A., 93
Unemployment, and jobs unfilled, 25
Unemployment, and robotics:
  and birth rate, 98
  new workers, scarcity of, 98, 99
  service sector, 99
Unimation, 27, 28, 55
University of Chicago Magazine, 74
United States:
  education, 48
  productivity decline in, 5
  robotics in, 48–50
Unvaryingness, of robot work, 41

V

Versatility, lack of in automation, 11

Vision, artificial:
  acts for robot to do, 71
  approximations of patterns, 69
  and bands outside human reach, 66
  edges, scanning for, 68
  fields of view, 69, 70
  object discrimination, 67
  partial views of objects, 71
  perimeters, storing of, 69
  photocells, 67
  profiles, recognition of, 68
  silhouette vision, 68
  and tactile sense, 66
Vision, human:
  and boundaries in image, 61
  and color, 60–61
  data from retina, 61
  data load reduction, 61, 62
  extrapolation by brain, 61–62
  mental models, connections of, 64
  motion of eye, 62
  optical illusions, 63
  recognition thresholds, 63, 64
  retina, receptors in, 61
  vertical lines, 63
Von Tiesenhausen, Georg, 7

W

Wages, increase of, trends, 101
Waltz, David, 75
Weizenbaum, Joseph, 73, 74, 109
Welfare system, growth of, 81–82
What is Life?, 81
Wilson, Richard C., 110
Winkless, Nels, 116
Woodrow, Jim, 107

Y

Yamazaki, Machinery Works, 20
Year of the Robot, The, 115–16